The Institute of Biology aims to advance both the science and practice of biology. Besides providing the general editors for this series, the Institute publishes two journals *Biologist* and the *Journal of Biological Education*, conducts examinations, arranges national and local meetings and represents the views of its members to government and other bodies. The emphasis of the *Studies in Biology* will be on subjects covering major parts of first-year undergraduate courses. We will be publishing new editions of the 'bestsellers' as well as publishing additional new titles.

Titles available in this series

An Introduction to Genetic Engineering, D. S. T. Nicholl

Photosynthesis, 6th edition, D. O. Hall and K. K. Rao

Introductory Microbiology, J. Heritage, E. G. V. Evans and R. A. Killington

Biotechnology, 3rd edition, J. E. Smith

An Introduction to Parasitology, Bernard E. Matthews

Essentials of Animal Behaviour, P. J. B. Slater

Microbiology in Action, J. Heritage, E. G. V. Evans and R. A. Killington

Microbiology in Action

Microbes play an important role in our everyday lives. As agents of infectious disease they cause untold human misery, yet their beneficial activities are manifold, ranging from the natural cycling of chemical elements through to the production of food, beverages and pharmaceuticals. In this introductory level text, the authors provide a clear and accessible account of the interactions between microbes, their environment and other organisms, using examples of both beneficial and adverse activities. The book begins by considering beneficial activities, focusing on environmental microbiology and manufacturing, and then moves on to consider some of the more adverse aspects, particularly the myriad of diseases to which we are susceptible and the treatments currently in use.

This book is the companion volume to *Introductory Microbiology*, also published in this series. It provides essential reading for biological science and medical undergraduates, as well as being of interest to sixth form students and their teachers.

JOHN HERITAGE is a Senior Lecturer in Microbiology at the University of Leeds where his research interests centre on the evolution and dissemination of antibiotic-resistance determinants in Gram-negative bacteria. He is a member of the UK Government Advisory Committee on Novel Foods and Processes.

GLYN EVANS is Professor of Medical Mycology at the University of Leeds and Head of a UK Public Health Laboratory Service Mycology Reference Laboratory. His research interests concern aspects of epidemiology, serodiagnosis, treatment and pathogenesis of fungal infections.

DICK KILLINGTON is a Senior Lecturer in Microbiology at the University of Leeds where his research focuses on biochemical and immunological aspects of herpesviruses, hepatitis C virus and rhinoviruses.

Microbiology in action

J. Heritage, E. G. V. Evans and R. A. Killington
Department of Microbiology, University of Leeds

Published in association with the Institute of Biology

CAMBRIDGE
UNIVERSITY PRESS

PUBLISHED BY THE PRESS SYNDICATE OF THE UNIVERSITY OF CAMBRIDGE
The Pitt Building, Trumpington Street, Cambridge, United Kingdom

CAMBRIDGE UNIVERSITY PRESS
The Edinburgh Building, Cambridge CB2 2RU, UK   http://www.cup.cam.ac.uk
40 West 20th Street, New York, NY 10011-4211, USA   http://www.cup.org
10 Stamford Road, Oakleigh, Melbourne 3166, Australia

First published 1999

Printed in the United Kingdom at the University Press, Cambridge

Typeset in Monotype Garamond 11/13 in QuarkXPress™   [SE]

A catalogue record for this book is available from the British Library

Library of Congress Cataloguing in Publication data

Heritage J.
Microbiology in action / J. Heritage, E. G. V. Evans, and R. A.
Killington.
    p.   cm.
Includes bibliographical references and index.
ISBN 0 521 62111 9. – ISBN 0 521 62912 8 (pbk.)

1. Microbiology.   I. Evans, E. G. V. (Emlyn Glyn Vaughn)
II. Killington, R. A.   III. Title.
QR41.2.H463   1999
579–dc21   98-44695 CIP

ISBN 0 521 62111 9 hardback
ISBN 0 521 62912 8 paperback

Contents

Preface

When we wrote *Introductory Microbiology* some very hard decisions had to be made concerning the contents of the book. We were constrained by the style of the *Studies in Biology* series to write a book of no more than 200 pages. In the end, we decided that students needed a description of what microorganisms were and how they can be safely manipulated before appreciating what they can do. We therefore took the decision to base our first book on these fundamental aspects of the subject. We were convinced at the time, however, that we could fill a second book with the material that we had omitted from the first. All we had to do was to persuade a publisher that students need to know about much more than we could include in that book.

Tim Benton, who edited our *Introductory Microbiology*, was so pleased with our proposal that he accepted our ideas and then promptly left Cambridge University Press to take up an academic career. We are not suggesting that this career change has any bearing on Tim's ability to make rational decisions or on the viability of our proposals. The project was handed on to Barnaby Willitts. He was very supportive throughout the writing of this book. As the deadline for submission arose, however, Barnaby left the press (and the country). The project was then handed to Maria Murphy. We owe all those who played a part in producing this book a debt of thanks.

The title chosen for this book is *Microbes in Action*. This implies that microbes have an active impact on our lives. We have framed the text around a series of questions. The answers to these questions illustrate the effects microbes have on humans. In planning this book, we hope to show that microorganisms are more than just the agents of infectious diseases. Without the activities of microbes, for example in the biological cycling of chemical elements, life as we know it would very soon become extinct. To indicate the importance of such beneficial processes, environmental microbiology and the role of microbes in manufacturing have been placed at the beginning of the book. Microbes do, however, cause untold human misery as well as bringing unnoticed benefits. Very early on we reveal how microbes nearly caused the downfall of Winchester Cathedral. This fulfils the promise we made in the preface to *Introductory Microbiology*, even if we mistook the wood used to build its raft. The cathedral was built on a beech raft, not one of oak. Furthermore, all three authors have research interests that lie within the sphere of medical microbiology. It is for these reasons that the majority of the text describes how microbes harm humans and how we can control them. We trust that our readers will forgive our bias in that direction.

We complained in the preface to *Introductory Microbiology* that there was insufficient room to cover all of microbiology in a text of that size. We have again failed to include everything of interest that we had to omit from our first book. It would be churlish to complain again about the lack of space. We have, however, left uncovered those things which we ought to have covered ... And there is no health in us. To get around this problem we have included a list of texts through which interested readers may extend their knowledge. We hope that this provides recompense for our manifold sins and wickedness.

Constraints of space have beaten us once again. We have concentrated on the areas covered by our research interests. This is why the book is largely devoted to **bacteria**, **fungi** and **viruses**. We have had to omit important material on parasites, for example, although books listed in the *Further reading* section should cover the material that we have left out. If readers are wondering about the differences between bacteria, fungi and viruses, or how **prokaryotic** cells differs from **eukaryotic** cells, then we can do no better than recommend *Introductory Microbiology*. Alternatively, the reader could always refer to the glossary at the back of this book.

During the production of this book, my wife took our children to visit her mother for two weeks while the writing was at its most difficult. This was a time when too much of the book had been written to cancel the project and not enough had been assembled to allow sight of the end of the writing. I am truly grateful to my family for the break this gave me to write uninterrupted. Without this gesture it is doubtful that you would now be holding this book in your hand. I owe a huge debt to my family. They showed great patience during the writing of this book.

Again, this book would not have been possible without the assistance of colleagues who have advised on different aspects of the project. Our thanks are extended to those who have participated in the production of this book but the mistakes that are left remain our responsibility.

We were pleased with the success of *Introductory Microbiology*: Cambridge University Press must have been similarly pleased to allow us to finish this project. We are grateful for that trust and hope that you will enjoy this book as much as its predecessor.

JH
York

1

The microbiology of soil and of nutrient cycling

Soil is a dynamic habitat for an enormous variety of life-forms. It gives a mechanical support to plants from which they extract nutrients. It shelters many animal types, from invertebrates such as worms and insects up to mammals like rabbits, moles, foxes and badgers. It also provides habitats colonised by a staggering variety of microorganisms. All these forms of life interact with one another and with the soil to create continually changing conditions. This allows an on-going evolution of soil habitats.

The activity of living organisms in soil helps to control its quality, depth, structure and properties. The climate, slope, locale and bedrock also contribute to the nature of soil in different locations. The interactions between these multiple factors are responsible for the variation of soil types. Consequently, the same fundamental soil structure in different locations may be found to support very different biological communities. These complex communities contribute significantly to the continuous cycling of nutrients across the globe.

1.1   What habitats are provided by soil?

Soil forms by the breakdown of bedrock material. Erosion of rocks may be the result of chemical, physical or biological activity, or combinations of the three factors. Dissolved carbon dioxide and other gases cause rain water to become slightly to moderately acid. This pH effect may cause the breakdown of rocks such as limestone. Physical or mechanical erosion can result from the action of wind or water, including ice erosion. The growth of plant roots and the digging or burrowing activities of animals contribute to the mechanical breakdown of soil. Microbial activity by **thermoacidophilic** bacteria, such as those found in coal slag heaps, results in an extremely acid environment. Leaching of acid from slag heaps may cause chemical changes in bedrock.

Naked rocks provide a very inhospitable habitat. Even these may, however, be colonised. There is evidence for colonisation all around us. Next time you visit a graveyard, look for lichens on the headstones. Lichens are microbial colonisers of rocks. This is true even if the rock is not in its original environment. Gravestones are conveniently dated. By comparing the age of different headstones and the degree of colonisation you can get some idea of the time it takes to colonise native rocks.

Among the first rock colonisers are cyanobacteria. Parent rocks do not provide nitrogen in a form that is readily available for biological systems. Bacteria are unique among life-forms in that they can fix atmospheric nitrogen so that it can be used by other organisms. Cyanobacteria are ideally placed to colonise rock surfaces because they are nitrogen-fixing **photolithotrophs**. They require only light and inorganic nutrients to grow. Cyanobacteria can provide both fixed nitrogen and carbon compounds that can be used as nutrients by other organisms. They are responsible for the initial deposition of organic matter on exposed rocks. This initiates the biological processes that lead to soil formation and to nutrient cycling. The colonisation of rocks by cyanobacteria is the first step in the transformation of naked rock into soil suitable for the support of plant and animal life. The microbes present in the soil are responsible for re-cycling organic and inorganic material and play an important part in the dynamic regeneration of soil.

As soils develop and evolve, the smallest particles are found nearest the surface of the ground and particle size increases steadily down to the bedrock. Soil particles may be classified by size (Fig. 1.1). Sand particles are typically between 50 micrometres and 2 millimetres. Silt particles are smaller than sand particles, being between 2 micrometres and 50 micrometres. Clay particles are smaller than 2 micrometres. The sizes of the particles present in soil profoundly affect its nature. One cubic metre of sand may contain approximately 10^8 particles and has a surface area of about 6000 square metres. The same volume of clay may contain 10^{17} particles with a surface area of about 6 million square metres. As the size of particle decreases, the number of particles present in a unit volume of soil increases exponentially, as does the surface area of the soil. This has important consequences for water retention and hence for other properties of the soil.

Sandy soils, with their relatively small surface area, cannot retain water very well and drain very quickly. This may lead to the formation of arid soils. At

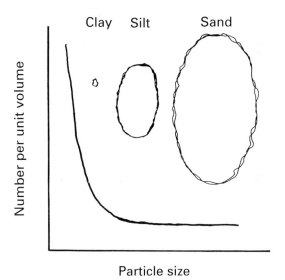

Fig. 1.1. The relative sizes of soil particles. The clay particle is small, the silt particle is of average size and the sand particle is large.

the other extreme, clays have a very large surface area and retain water very easily. Clays also tend not to be porous. As a result of water retention, they also tend to form anaerobic environments. Neither extreme provides an ideal habitat, other than for specialised life-forms. The most fertile soils are loams. These contain a mixture of sand, silt and clay particles and provide a diversity of microhabitats capable of supporting a wide range of organisms. These organisms interact to modify the atmosphere between particles of soil. Consequently, the atmosphere within the soil differs from that above ground. Microbial and other metabolisms use some of the available oxygen present in the soil and so there is less oxygen beneath the ground than there is in the air above the soil surface. Similarly, carbon dioxide is generated as a by-product of microbial metabolism and there is a higher concentration of carbon dioxide within soil than above ground.

Soils may also be grouped by their organic content. At one extreme are mineral soils that have little or no organic content. Such soils are typical of desert environments. At the other extreme are bogs. There is a gradation of soil types between that found in deserts and that in bogs, with an ever-increasing organic content.

Plants are the major producers of organic material to be found in soil, and plant matter accumulates as litter. Animal faeces and the decomposing bodies of dead animals complement this organic supply. Artificially added fertilisers, herbicides and pesticides all affect the biological component and hence the organic content of soils. Horse dung and chicken manure are beloved of gardeners. Microbes play a central role in re-cycling such material. Besides re-cycling of naturally occurring organic compounds, soil microbes are responsible for the chemical degradation of pesticides. Not all pesticides are easily broken down, however. Those compounds that resist microbial decomposition and that consequently accumulate in the environment are known as recalcitrant pesticides.

During the evolution of a soil habitat, its organic content may eventually become predominant. The ultimate organic soil is found in a bog. Bogs are waterlogged and consequently form an anaerobic environment. Any dissolved oxygen is quickly used up by **facultative organisms**. This provides a very inhospitable environment for fungi and **aerobic bacteria**. Since these organisms tend to be responsible for the decomposition of organic structures, bogs provide excellent sites for the preservation of organic matter.

A striking example of the preservative effect of bogs is afforded by the existence of intact human bodies conserved for thousands of years. 'Pete Marsh' was one such specimen. His body was found in a bog in Cheshire. He was in such a good state of preservation that a forensic *post mortem* examination was possible on this archaeological find, showing that the man had died after being garrotted. 'Lindow Man', as he is also known, is now on view in a special atmospherically controlled chamber in the British Museum.

Winchester Cathedral was built on a peat bog. To support this magnificent structure, the medieval architects and masons raised the building on a huge raft made from beech trees. This raft provided a floating foundation for the Cathedral. The wood survived intact for hundreds of years, preserved by the anaerobic, waterlogged environment provided by the marshy ground upon which it rested. It was only during the early twentieth century that a crisis arose. The surrounding water-meadows were drained to conform to the agricultural practices then in fashion. The water table around the Cathedral started to fluctuate and the beech raft was exposed to the air for the first time in centuries. It was also exposed to the microorganisms responsible for wood decay. It was only owing to the engineering expertise of a single diver, William Walker, that the whole structure was saved from disaster. He spent years working alone under the cathedral underpinning its structure. A similar drop in the water table in the Black Bay area of Boston has caused considerable problems of subsidence in some of the older buildings in the area. Again, this is caused by oxygen-dependent fungi rotting the previously soaked timber piles on which the buildings were erected.

Soils contain many aerobic and facultative organisms and, because of the microbial manipulation of microenvironments, soils may harbour a large number of **obligate anaerobes**. Bacteria are the largest group of soil microbes, both in total number and in diversity. Indeed the presence of bacteria gives freshly dug soil its characteristic 'earthy' smell. The odour is that of **geosmin**, a secondary metabolite produced by streptomycete bacteria.

Microscopic examination of soil reveals vast numbers of bacteria are present. Typically there are about 10^8 to 10^9 per gram dry weight of soil. Only a tiny fraction of these can be cultivated upon laboratory culture media. Scientists have yet to provide appropriate culture conditions for the vast majority of soil microbes. Many live in complex communities in which individuals cross-feed one another in a manner that cannot be replicated when the microbes are placed in artificial culture. The microbial activity of soil is severely underestimated using artificial culture. An estimate of the microbial activity of soil is further influenced by the fact that many soil bacteria and fungi are present as dormant spores. These may germinate when brought into contact with a rich artificial growth medium. Spores may also flourish when introduced into cuts and grazes. Gardeners are particularly prone to tetanus, when spores of *Clostridium tetani* are introduced into minor trauma sites.

For many years, the study of soil microbiology was severely limited because of our inability to cultivate the vast majority of soil microbes in artificial culture. Today, great advances are being made by the application of molecular biological techniques to this problem. Sensitive isotope studies are yielding information on the metabolism of soil microbes and **polymerase chain reaction** (PCR) technology is being used to study the taxonomy of non-cultivable bacteria, particularly exploiting 16S ribosomal RNA (rRNA) structure. The structure of the 16S rRNA is conserved within members of a species whereas different species show divergent 16S rRNA structures. Therefore this provides a very useful target in taxonomic studies.

Both bacteria and fungi provide an abundant source of food for soil protozoa. The most commonly encountered soil protozoa include flagellates and amoebas. The abundance of such creatures depends upon the quantity and type of organic matter present in the soil sample. Protozoa play a key role in the regulation and maintenance of the equilibrium of soil microbes. Whereas many microbes obtain their nutrients from solution, protozoa are frequently found to be of a scavenging nature, obtaining their nutrients by devouring other microbes.

The distribution of microbes throughout the soil is not even. Microorganisms tend to cluster around the roots of higher plants. This phenomenon is referred to as the **rhizosphere effect** (*rhiza*: Greek for root; hence the rhizosphere is the region surrounding the roots of a plant). The majority of microorganisms found in the rhizosphere are bacteria, but fungi and proto-zoa also congregate in this region. Microorganisms are thought to gain nutri-ents from plants, and **auxotrophic** mutants requiring various amino acids have been isolated from the rhizosphere. Plants may also derive benefit from this arrangement. Bacteria may fix nitrogen in a form that can be taken up and used by plants. In certain circumstances, the association between microorgan-isms and higher plants can become very intimate. **Mycorrhizas** are formed when roots become intimately associated with fungi. Root nodules provide another important example of the close association between leguminous plants and nitrogen-fixing bacteria. In this instance bacteria rather than fungi are involved in the association with plants.

1.2　How are microbes involved in nutrient cycling?

Life on Earth is based on carbon. Water and simple organic compounds such as carbon dioxide become elaborated into complex, carbon-based organic structures. These compounds include other elements besides carbon, oxygen and hydrogen. Nitrogen is found in nucleic acids, amino acids and proteins. Phosphorous is a component of nucleic acids, lipids, energy storage com-pounds and other organic phosphates. Sulphur is found principally in certain amino acids and proteins. All of these elements are continuously cycled through the ecosystem. Many natural biological cycling processes require ele-ments to be in different chemical states in different stages of the cycle. Phosphorous is an exception. It is always taken up as inorganic phosphates. Once absorbed into living organisms, biochemical processes transform phosphorous into more complex forms.

Inorganic phosphates are very widely distributed in nature but are fre-quently present as insoluble salts. So, despite an apparently plentiful supply of phosphorous, phosphates often represent a **limiting nutrient** in natural ecosystems. This means that as supplies of phosphates run out, uncontrolled growth of organisms is prevented. Insoluble phosphates can be converted into soluble phosphates. This may be achieved by the activity of the acid prod-ucts of bacterial **fermentations**. These may then be taken up into bacteria. Soluble phosphates may also be added to the land artificially, either as plant fertilisers or as organophosphate pesticides. Phosphates are also used in the manufacture of many detergents. These chemicals can end up in rivers and lakes, artificially increasing the concentration of biologically accessible phos-phates. This permits the overgrowth of algae in affected waters, resulting in **algal blooms**. These can deprive other plants of light, thus killing them and destroying the natural ecology of the affected waters. Some algal blooms may also be toxic to animals.

Besides the cycling of non-metal elements, microorganisms have a role in the biochemical transformation of metal ions. Bacteria such as *Thiobacillus ferrooxidans* and iron bacteria of the genus *Gallionella* are capable of oxidising ferrous (Fe^{2+}) iron into ferric (Fe^{3+}) iron. Many bacteria can reduce small quantities of ferric iron to its ferrous state. There is also a group of iron-respiring bacteria that obtain their energy by **respiration**. They use ferric iron as an electron acceptor in place of oxygen. Magnetotactic bacteria, exemplified by *Aquaspirillum magnetotacticum*, can transform iron into its magnetic salt magnetite. These bacteria act as biological magnets. Bacteria are also important in the transformation of manganese ions, where similar reactions to those seen with iron are observed.

Without the cycling of elements, the continuation of life on Earth would be impossible, since essential nutrients would rapidly be taken up by organisms and locked in a form that cannot be used by others. The reactions involved in elemental cycling are often chemical in nature, but biochemical reactions also play an important part in the cycling of elements. Microbes are of prime importance in this process.

In a complete ecosystem, **photolithotrophs** or **chemolithotrophs** are found in association with **chemoorganotrophs** or **photoorganotrophs,** and nutrients continually cycle between these different types of organism. Lithotrophs gain energy from the metabolism of inorganic compounds such as carbon dioxide whereas organotrophs need a supply of complex organic molecules from which they derive energy. Phototrophs require light as a source of energy but chemotrophs can grow in the dark, obtaining their energy from chemical compounds. The rate of cycling of inorganic compounds has been estimated and different compounds cycle at very different rates. It is thought to take 2 million years for every molecule of water on the planet to be split as a result of photosynthesis and then to be regenerated by other life-forms. Photosynthesis may be mediated either by plants or photosynthetic microbes. The process of photosynthesis releases atmospheric oxygen. It is probable that all atmospheric oxygen is of biological origin and its cycling is thought to take about 2000 years. Photosynthesis is also responsible for the uptake of carbon dioxide into organic compounds. Carbon dioxide is released from these during respiration and some fermentations. It only takes about 300 years to cycle the atmospheric carbon dioxide.

Because of our familiarity with green plants, life without photosynthesis is perhaps difficult to imagine. This is, after all, the reaction that provides us with the oxygen that we need to survive. It should be remembered, however, that photosynthesis is responsible for the production of molecular oxygen. This element is highly toxic to many life-forms. Life on Earth evolved at a time when there was little or no oxygen in the atmosphere. Aerobic organisms can only survive because they have evolved elaborate protection mechanisms to limit the toxicity of oxygen. Equally, not all life depends on sunlight. In the dark depths of both the Atlantic and Pacific oceans are thermal vents in the Earth's crust. These provide a source of heat and chemical energy that chemolithotrophic bacteria can use. In turn, these bacteria provide a food source for a range of invertebrates. These rich and diverse communities spend their entire lives in pitch darkness around the 'black smokers'.

1.2.1   How is carbon cycled?

Most people are familiar with the aerobic carbon cycle. During photosynthesis, organic compounds are generated as a result of the fixation of carbon dioxide. Photosynthetic plants and microbes are the primary producers of organic carbon compounds and these provide nutrients for other organisms. These organisms act as consumers of organic carbon and break down organic material in the processes of fermentation and respiration. Chemoorganotrophic microbes break down organic carbon compounds to release carbon dioxide. Chemolithotrophic bacteria can assimilate inorganic carbon into organic matter in the dark. Certain bacteria are also capable of anaerobic carbon cycling. Fermentation reactions, common in bacteria that are found in water and anaerobic soils, are responsible for the breakdown of organic chemicals into carbon dioxide or methane. Hydrogen gas may be released as a product of some fermentations. Methane can itself act as a carbon and energy source for methane-oxidising bacteria. These bacteria can generate sugars and amino acids from methane found in their environments, again helping with the cycling of carbon compounds.

1.2.2   How is nitrogen cycled?

One of the crucial steps in the advancement of human civilisation was the development of agriculture. This involves the artificial manipulation of the natural environment to maximise the yield of food crops and livestock. With the development of agriculture came the need to maximise the fertility of soils. The availability of fixed nitrogen in a form that can be used by crop plants is of prime importance in determining the fertility of soil. As a consequence, the biological nitrogen cycle (see Fig. 1.4 below) is of fundamental importance, both to agriculture and to natural ecology.

Inorganic nitrogen compounds such as nitrates, nitrites and ammonia are converted into organic nitrogen compounds such as proteins and nucleic acids in the process of **nitrogen assimilation**. Many bacteria reduce nitrates to nitrites and some bacteria further reduce nitrites to ammonia. Ammonium salts may then be incorporated into organic polymers in the process of **assimilatory nitrate reduction**. Ammonia is primarily fixed into organic matter by way of amino acids such as glutamate and glutamine. Other nitrogen compounds can be made from these.

For the continued cycling of nitrogen, organic nitrogen compounds must be broken down to release ammonia. **Putrefactive metabolism** yields considerable quantities of ammonia from **biopolymers** that contain nitrogen. Bacteria may also produce **urease**, an enzyme that breaks down urea to liberate carbon dioxide, water and ammonia. The quantity of ammonia released by the urease of *Helicobacter pylori* is sufficient to protect this bacterium from the acid pH of the human stomach.

Bacteria are also involved in the inorganic cycling of nitrogen compounds. **Nitrifying bacteria** are responsible for the biological oxidation of ammonia. These bacteria are chemolithotrophs, obtaining chemical energy from the oxidation process. This energy is used to elaborate organic compounds from carbon dioxide. Nitrifying bacteria such as those of the genus *Nitrosomonas* produce nitrite ions from the oxidation of ammonia. Bacteria of the genus *Nitrobacter* and a few other genera can oxidise nitrites to nitrates.

As well as their role in the nitrogen cycle, nitrifying bacteria may have a more sinister activity, as illustrated by their effects on buildings such as the cathedrals at Cologne and Regensburg. They have been shown to colonise the sandstone used to build these churches. Water carries the bacteria from the surface and into the matrix of the stone to a depth of up to five millimetres. Here they produce quantities of nitrous and nitric acid sufficient to cause erosion of the stone. Consequently, the decay of great public buildings may not be exclusively caused by acid rain generated by industrial pollution.

Nitrates may be used by some bacteria instead of oxygen for a type of respiration referred to as **dissimilatory nitrate reduction**. During this process, nitrate is reduced to nitrite and thence to ammonia. This may then be assimilated into organic compounds as described above. Not all bacteria follow this pathway, however. Bacteria of the genus *Pseudomonas,* micrococci and *Thiobacillus* species can reduce nitrates to liberate nitrogen gas into the environment. Bacteria that can generate nitrogen gas from the reduction of nitrates are commonly found in organically rich soils, compost heaps and in sewage treatment plants.

To complete the inorganic nitrogen cycle, nitrogen gas must be fixed in a form that can be used by living organisms. If this were not the case, life on Earth would only have continued until all the available nitrogen compounds had been converted into nitrogen gas. Nitrogen is converted into ammonia in the process of **nitrogen fixation.** Bacteria are the only life-forms capable of the biological fixation of nitrogen. They are of vital importance if life is to continue on this planet. Green plants are the main producers of organic matter in the biosphere and they require a supply of fixed nitrogen for this process. Fixed nitrogen may be obtained through the death and lysis of free-living nitrogen-fixing bacteria. Nitrogen-fixing bacteria, however, frequently form close associations with plants. In some cases, the relationship becomes so intimate that bacteria live as **endosymbionts** within plant tissues. Bacteria supply the plant with all of its fixed nitrogen demands. In return, they receive a supply of organic carbon compounds (Fig. 1.2).

Not all nitrogen fixation occurs as a result of biological processes. Nitrogenous fertilisers are produced in vast quantities by the agrochemical industry. Oxides of nitrogen are also produced by natural and artificial phenomena in the environment. Ultraviolet irradiation and lightning facilitate the oxidation of nitrogen, particularly in the upper atmosphere. At ground level, these reactions are augmented by electrical discharges and in particular by the activity of car engines.

Biological nitrogen fixation is catalysed by **nitrogenase**. The activity of this enzyme is rapidly lost in the presence of oxygen. Nitrogen-fixing bacteria have evolved a number of strategies for protecting nitrogenase from the harmful effects of oxygen. The simplest solution is to grow under anaerobic conditions. Nitrogen fixers such as *Clostridium pasteurianum* and *Desulfovibrio desulfuricans* are obligate anaerobes. In consequence, they can only grow under conditions that protect the activity of nitrogenase.

Many bacteria are facultative anaerobes. Nitrogen-fixing facultative bacteria are generally only capable of fixing nitrogen when they are growing in anaerobic environments. Examples of such bacteria include some of the Enterobacteriaceae, such as *Enterobacter* species, as well as facultative members of the genus *Bacillus*.

The Gram-negative bacterium *Klebsiella pneumoniae* is capable of nitrogen fixation in a microaerophilic atmosphere as well as in anaerobic conditions. This is because in a microaerophilic atmosphere bacterial respiration can effectively reduce the local oxygen tension to zero. This permits nitrogenase to function. This provides an example of respiration fulfilling two demands

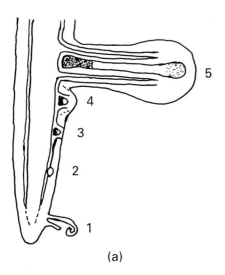

(a)

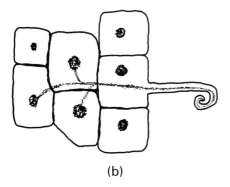

(b)

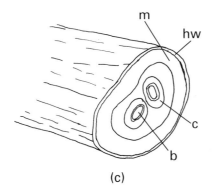

(c)

Fig. 1.2. The development of a root nodule. (*a*) Macroscopic view of the nodule. The first stage sees the root hairs curling (1). An infection thread develops and stimulates the root cortical cells to divide. A nodule meristem then begins to develop (2). Bacteriods increase within the developing nodule (3) which then emerges from the root (4). In the fully developed nodule (5), the region furthest from the root is the region that is newly colonised with bacteria. Nitrogen fixation occurs in the middle section of the root and the region nearest the root represents a senescent area. (*b*) The infection thread enters the root hair at its curled tip. It then grows down the hair and through the epidermal layer when it branches. Each branch becomes associated with a cortical cell nucleus. (c) Detail of the infection thread shows bacteroids (b) at the centre of the structure surrounded by bacterial capsular material (c). Beyond lies the thread matrix (m), material of bacterial origin. This is then surrounded by the host cell enclosed within its cell wall (hw).

v   h   a   h   th

Fig. 1.3. Differentiation of cells of a cyanobacterium. At either end of the filament are terminal cells. On the right is a heterocyst (th) as shown by the dark 'polar body' seen where it joins onto the filament. It is not common to see heterocysts as terminal cells. Vegetative cells (v) are capable of photosynthesis and have a blue-green coloration. These are the sites of carbon fixation. Regularly dispersed through the filament are heterocysts (h). It is in these non-photosynthetic cells that nitrogen fixation occurs. Heterocysts have thicker cell walls than vegetative cells and each has its polar bodies. The large granular cells are known as akinetes (a) and are produced as resting cells.

for a bacterium. The klebsiella cells obtain energy as ATP from respiration, while at the same time protecting nitrogenase. Truly microaerophilic bacteria have also been found that fix nitrogen. Such bacteria are common in soils and other specialised habitats where oxygen does not penetrate well.

Many aerobic nitrogen-fixing bacteria such as the free-living forms of *Rhizobium* species live as microaerophiles when deprived of a source of fixed nitrogen. Reduction of oxygen tension in microaerophilic conditions is achieved by bacterial respiration. In certain species within the genus *Azotobacter*, respiration is sufficiently active to allow bacteria to fix nitrogen, even under aerobic conditions. This requires an extremely high respiration rate. Consequently, aerobic nitrogen fixation by azotobacters can only take place when the bacteria have a plentiful supply of organic carbon compounds to act as substrates for respiration.

Cyanobacteria have adopted an alternative strategy for the protection of nitrogenase. They undergo cellular differentiation, with nitrogen fixation being confined to specialised cells known as **heterocysts**. These develop in response to nitrogen starvation of fixed sources of nitrogen. In free-living cyanobacteria, heterocysts account for less than 10% of the filament cells, but when cyanobacteria are found in nitrogen-fixing symbioses, the frequency of heterocysts in filaments rises dramatically. Once the cyanobacterial partner is isolated from such symbioses, the heterocyst frequency falls again. Heterocysts function to exclude oxygen through ultrastructural and metabolic changes to the cell (Fig. 1.3). The principal ultrastructural modification is the synthesis of three extra layers of cell wall material around the mature hetero-cyst. These extra layers help to prevent the diffusion of oxygen into the cell. Heterocysts also fail to produce phycocyanin, a light-harvesting pigment that gives cyanobacteria their typical blue–green appearance. As a consequence, heterocysts appear greener and paler than vegetative cells in the cyanobacterial filament. Phycocyanin plays an important role in the generation of oxygen during photosynthesis. The absence of phycocyanin prevents photosynthetic oxygen formation within the heterocyst. Metabolic functions within the heterocysts are also modified. In this respect, heterocysts may be described as anaerobic islands within aerobic filaments.

1.2.3 How is sulphur cycled?

Sulphur is the substance of brimstone. Anyone who has visited the sulphur springs in volcanically active areas and who has experienced the choking sulphurous fumes would hardly credit that this element was compatible with life. Sulphur is, however, a minor but important component of proteins. It is the disulphide bridges that give many proteins their active three-dimensional structure. The biological cycling of sulphur is, in many respects, similar to that of the nitrogen cycle (Fig. 1.4). It is of a lesser economic importance, however. Consequently, the processes have not been studied in such great detail as those involved in cycling of nitrogen through the biosphere.

Unlike the nitrogen cycle, evidence for the biological sulphur cycle can be gained by a simple visit to the seaside. The sand on many beaches around the UK is rich in organic matter. This is especially true where sewage is discharged into the sea. The determined builder of sand castles will probably notice that below the surface lies a black layer of sand. In this region, sulphate-reducing bacteria act upon the sulphur compounds in the accumulated organic matter, releasing hydrogen sulphide. This, in turn, reacts with iron in the sand, and in the wet, anaerobic conditions under the surface of the sand, black iron sulphide is formed. If this black sand is added to the top of a sand castle it will almost miraculously revert to the colour of the native sand. This happens as the iron sulphide is broken down on exposure to the air to produce iron oxides.

As with nitrogen, plants and animals are unable to use the elemental form of sulphur. *Thiobacillus thiooxidans* can, however, produce sulphates as a result of the biological oxidation of elemental sulphur. It is as inorganic sulphates that most bacteria assimilate sulphur. Sulphates are assimilated into organic compounds by reduction to hydrogen sulphide. This is then incorporated into the amino acid cysteine by reaction with *O*-acetylserine. Cysteine is then further metabolised to generate other organic sulphur compounds.

The purple and green sulphur bacteria can use reduced sulphur compounds such as hydrogen sulphide as electron donors for their photosynthetic metabolism. As hydrogen sulphide is used, sulphur granules are generated. It

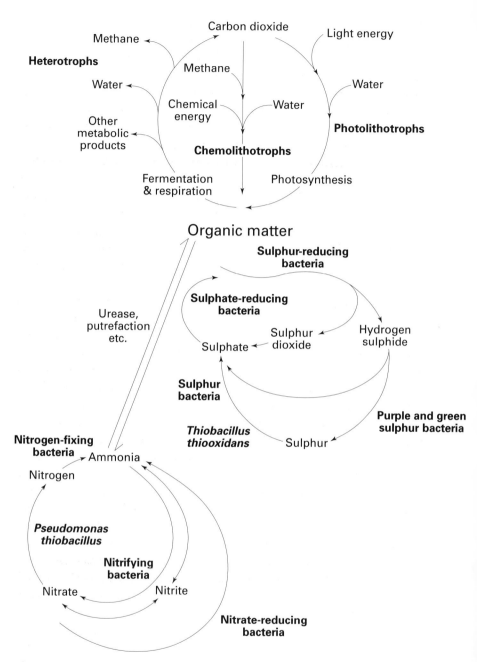

Fig. 1.4. The role of microbes in the cycling of carbon nitrogen and sulphur.

has been proposed that the geological deposits of elemental sulphur found in rocks around the world are **biogenic** in origin. This means that they are derived from the metabolic activity of organisms, particularly the photosynthetic sulphur bacteria that lived in ancient oceans.

The putrefactive metabolism of protein typical of certain **anaerobic bacteria** results in the liberation of large quantities of hydrogen sulphide. It is surely no coincidence that hydrogen sulphide has always been identified with the smell of rotten eggs. This provides just one example of **dissimilatory sulphur metabolism** – a diversity of bacteria can reduce sulphur to produce sulphides. Some microbes are capable of metabolising sulphates or elemental sulphur as electron acceptors in the process of anaerobic respiration. Dissimilatory sulphur- or sulphate-reducing bacteria are found in a wide array of anaerobic environments. These include anaerobic muds, freshwater sediments, stagnant waters rich in organic matter, sewage treatment plants and the intestines of animals and humans. This may lead to emission of smells reminiscent of rotten eggs.

Bacteria of the genus *Desulfovibrio* may be of economic importance in both the oil industry and in agriculture. They are also a common cause of the blackening of anaerobic muds as a result of the generation of sulphides. The sulphides produced by desulfovibrios can be a direct cause of the corrosion of iron pipes that are buried in the ground. Desulfovibrios also produce enzymes that greatly enhance the corrosion of iron. This is a major problem for the oil industry, since it makes extensive use of sub-terranean iron pipes.

In well-aerated soil, desulfovibrios are not a major component of the ecosystem. They can accumulate to very high numbers in the anaerobic soils of rice paddies, however. The hydrogen sulphide that they produce may have a significant inhibitory effect upon root development of the growing rice plants and this can have consequent devastating effect upon crop yields. However, not all the effects of microbial sulphur metabolism are economically or socially disastrous. Sulphur-metabolising bacteria have been harnessed because of their potentially beneficial effects. For example, sulphur is removed from coal before burning by the activity of *Thiobacillus thiooxidans*. This helps to reduce acid pollution in the atmosphere when the coal is burned.

Archaebacteria may also be capable of sulphur metabolism. Thermoplasmas live in coal slag heaps where they generate large amounts of sulphuric acid. Members of the genus *Sulfolobus* are found in hot sulphur springs such as those found in volcanic areas. Like the thermoplasmas, these bacteria are **thermophiles** and they can metabolise hydrogen sulphide or elemental sulphur to produce sulphuric acid.

2

Plant–microbe interactions

2.1   What are mycorrhizas?

The name mycorrhiza is derived from two Greek words, *mukes* meaning a mushroom and *rhiza*, a root, illustrating a very important mutualistic interrelationship between plants and fungi. These partnerships have a long history. The fossil record shows that fungi and higher plants have lived in the close association of mycorrhizal relationships for at least 400 million years. The first recorded observations of mycorrhizal associations were made in the mid-nineteenth century. The numbers of plants that form mutualistic associations with fungi perhaps best illustrates the importance of mycorrhizal relationships. Over 80% of higher plants and ferns grow in association with a fungal partner. The range of higher plants affected include hard- and softwood trees, shrubs and other flowering plants as well as grasses.

The fungi that form mycorrhizal partnerships greatly extend the active surface area of the root system of plants. Fungi replace and extend the root system of the plant. The roots of trees that carry mycorrhizas are typically short and **dichotomously** branched. Unaffected roots are much longer. Orchids have evolved to such a degree that their mycorrhizal fungi have even replaced the plant root hairs. Plants that support mycorrhizas have much greater access to inorganic nutrients, particularly nitrates, phosphates and water. In return, the plant partner supplies its fungus with a source of organic nutrients and, in many cases, **vitamins**. Supplying organic matter and vitamins to their fungal partner incurs energy costs for the plant yet this seems to be compensated for by the ability of the fungus to provide its partner with *its* nutrients.

It is not just the plant and the fungus that gain potential benefit from mycorrhizal associations. The extensive fungal **mycelia** found in mycorrhizas are very important factors in maintaining soil structure since they can help to bind soil particles together. This will assist in preventing soil erosion and loss. The presence of fungi in poor-quality soil will greatly increase the chances of plants being able to thrive in such locations upon the formation of mycorrhizal relationships. Fungi that can form mycorrhizal associations can make the difference between a marginal and a fertile habitat. The activity of bacteria within the biosphere around plant roots may also play an important role in the successful establishment of a mycorrhizal partnership. Early attempts at re-forestation in locations such as Puerto Rico were largely disappointing, but the success of such schemes has markedly improved since fungal spores were incorporated into the planting programmes. Similarly, prairie landscape can be reclaimed when planting is accompanied by the re-introduction of fungi.

The nutritional advantages that each partner derives in a mycorrhizal association provide a possible explanation for the success of such relationships, but the partners can derive other benefits. Some mycorrhizal fungi have been shown to produce auxins: plant hormones that stimulate growth. Fungi in mycorrhizal partnerships are also frequently found to produce antibiotics. These help to regulate the microenvironment around the plant roots and can play a role in the prevention of plant infection. Mycorrhizal fungi have been shown experimentally to provide protection against *Phytophthora infestans*, the fungus that causes potato blight. Given the benefits that mycorrhizal associations can bestow, it is perhaps ironic that certain fungi that form mutualistic mycorrhizal relationships with particular plants can cause considerable damage to other plants. Fungi of the genus *Pythium* typically interact with onion and lettuce plants. Such interactions are diverse. On occasions, the fungus is unable to colonise the plant. Sometimes, mycorrhizal associations are established when, under different conditions, these fungi can act as aggressive **pathogens**. The nature of the interaction between the fungus and its host in such conditions seems only to be regulated by the soil condition at the time.

This serves to illustrate the complex nature of biological partnerships. It also indicates that plants require mechanisms to prevent potential damage caused by their fungal partners. These mechanisms are little understood at present. It is known, however, that plants do have a variety of defences against infection and that these may be moderated in specimens that have mycorrhizal relationships. The phenolic compounds and certain protective proteins produced by plant roots can have a significant antimicrobial effect. Plant specimens that grow without a fungal partner produce significantly more of these compounds than do mycorrhizal plants. Mycorrhizal hosts do, however, produce chitinases and peroxidases. It is thought that these play a significant role in preventing the fungal partner from becoming too invasive.

The fungi that form mycorrhizal associations with trees found in temperate woodlands are typically higher basidiomycete fungi. It is these fungi that create the familiar 'fairy rings' seen in woods. There is little specificity regarding these associations; various trees can form mycorrhizas with a given fungus and *vice versa*. More rarely, ascomycete fungi are capable of forming mycorrhizas. Perhaps the most famous example is the truffle, beloved of gastronomes. There are several varieties of truffle. The outer rind is formed from specialised mycelia and the body of the truffle comprises mycelia and fruiting bodies. In contrast to the mycorrhizas that typically form in woodland trees are the mycorrhizal associations formed by orchids. Different species of orchid each have a highly specific mycorrhizal relationship with its own particular fungus. The fungi that form mycorrhizas with orchids are most often lower fungi. Usually they are phycomycetes, typically zygmycotina. The degree of interdependence is illustrated by the observation that in many cases neither partner can be cultivated on its own. With orchids, the mutualistic relationship seen in mycorrhizas has evolved to form a true symbiosis. Indeed, many orchids cannot grow in sterile soil, even if the same species has grown successfully in the same soil sample before sterilisation.

There are various degrees of association between the fungus and its plant in a mycorrhizal relationship. In the most superficial of mycorrhizas, the fungus merely surrounds the root of a plant. There is no penetration of the plant tissues. These are known as **ectomycorrhizas** (Fig. 2.1). The fungal partner found in an ectomycorrhiza is typically a basidiomycete fungus. These are typically found around the roots of trees and shrubs. In other mycorrhizas, the **endomycorrhizas**, the fungal partner penetrates the roots, again to varying degrees (Fig. 2.1). Zygomycete fungi most often enter into endomycorrhizal relationships. In some cases, the root tissue is penetrated by its fungal partner; in others, the fungus penetrates right into the cells of its host's roots. Inside the host cells, fungi may form vesicles, but at the most extreme level of root penetration, fungal mycelia branch out within the cells of the host to form feathery structures known as **arbuscules**. The name is derived from the Latin *arbor* meaning a tree. These greatly increase the surface area over which fungus–host relationships can occur, maximising the interchange between a plant and its fungal partner. The presence of arbuscular structures is generally considered to indicate that the fungus–host relationship in that case has become truly symbiotic.

There is, however, an even more extreme example of fungal penetration than that seen with the arbuscular mycorrhizas. In ericaceous plants, fungi do

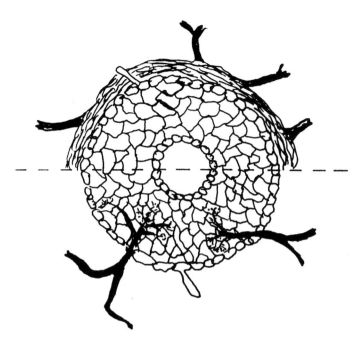

Fig. 2.1. Ecto- and endomycorrhizas. The upper portion of this figure represents a plant root that is shown surrounded by an ectomycorrhiza. Fungal mycelia surround the plant tissue without penetration. In the lower part of the diagram, mycelia penetrate the root cortex. The fungal cells of an endomycorrhiza are shown to invade their host's cells. Inside the host cell, each mycelium branches out to produce a feathery arbuscule to maximise the exchange of nutrients.

not simply penetrate the roots. Rather, the whole plant is infiltrated by fungal mycelia. This may explain the success of the heathers that grow in the harsh environments of locations such as the North Yorkshire Moors.

2.2   What symbioses do cyanobacteria form?

It is not just fungi that enter symbiotic relationships in the soil environment with plants. A variety of bacteria also act as symbionts. These bacteria often play a vital role in nutrient cycling and may be of great economic importance. Cyanobacteria are important examples of bacteria that form symbiotic relationships.

Cyanobacteria are an important group of microbes because they can fix carbon dioxide by photosynthesis. They also fix atmospheric nitrogen. Many organisms can assimilate inorganic carbon but very few can fix nitrogen.

Consequently, cyanobacteria are an important source of fixed nitrogen that can be used by other life-forms. They can be found in symbiotic association with a variety of organisms. In such arrangements, the carbon-fixing functions of the cyanobacterium are considerably reduced and they obtain much of their fixed carbon from their symbiotic partners. In turn, the symbiotic partner receives a rich supply of fixed nitrogen from the cyanobacterium. To achieve this, the cyanobacterium needs to devote much of its energies to nitrogen fixation. This it does in specialised cells known as heterocysts. The frequency of heterocysts in cyanobacteria found in symbioses is much greater than that seen in free-living forms. This high heterocyst frequency can be accomplished because the partner in a symbiotic relationship provides the cyanobacterium with fixed carbon compounds in return for a supply of fixed nitrogen.

Cyanobacteria commonly form associations with fungi, liverworts and water ferns, although they may associate with representatives in all phyla of higher plants and with certain animal species. The polar bears in the San Diego Zoo turned green at one time. This was because cyanobacteria were growing in the hollow hairs of the bear's fur. Polar bear fur is hollow to provide insulation, not a free home for cyanobacteria.

Most lichens consist of a fungus and a green alga. About 10% of lichens comprise a cyanobacterium living either together with a fungus as its sole partner or together with both a fungus and a green alga as a third symbiont. Consequently, lichens are the result of various symbiotic relationships between a fungus and other organisms. Fungi can fix neither carbon nor nitrogen and so in two-membered symbioses involving cyanobacteria the cyanobacterial partner must fix both carbon and nitrogen. The fungal partner supplies water and minerals to the cyanobacterium and protects its partner from excess light. In these lichens, the cyanobacterium must act as a phototroph and the heterocyst frequency is approximately the same as in the free-living form.

In three-membered lichens, the photosynthetic function is supplied by the green alga and the primary role of the cyanobacterium is to fix nitrogen. Consequently, the heterocyst frequency increases to about 20–30%. This compares with a typical heterocyst frequency of 4–8% in free-living cyanobacteria. Lichens are variable in the degree to which they can fix nitrogen. This can be illustrated by observing lichen growth on gravestones. The lichens that colonise vertical surfaces can be quite different from those that grow on the tops of these monuments. This is because the vertical surfaces do not collect bird droppings to the same extent as other surfaces. Bird droppings are an excellent source of fixed nitrogen and, therefore, lichens that grow in areas where bird droppings accumulate do not need to fix nitrogen as efficiently as those on the vertical surfaces.

There are five genera of liverworts that have been found to live in a symbiotic association with cyanobacteria. The cyanobacteria live in mucilaginous cavities found on the underside of the liverwort. To increase the area of contact between the partners, the liverwort grows branched filaments that extend through the cyanobacterial colony. This facilitates the exchange of nutrients between the two partners. Because the liverwort can provide the cyanobacterium with fixed carbon, the primary function of the cyanobacterium is to provide fixed nitrogen. The typical heterocyst frequency in such partnerships is between 30 and 40%.

Water ferns may also form symbioses with cyanobacteria, and one such living arrangement is of economic importance. The cyanobacterium *Anabaena azollae* lives inside the leaves of water ferns of the genus *Azolla*. This is an important source of fixed nitrogen in rice paddies, particularly in India and the Philippines. Elsewhere in the world, plants of the genus *Azolla* are considered as weeds, because of their habit of blocking waterways.

2.3 What symbioses do other nitrogen-fixing bacteria form?

Besides cyanobacteria, other nitrogen-fixing bacteria can form symbiotic associations with plants. The most familiar of these relationships is the formation of root nodules in leguminous plants. Nodules form in association with bacteria of the genus *Rhizobium*. Less well studied are the associations between actinomycete-like bacteria of the genus *Frankia* and non-leguminous plants.

Angiosperm plants of the family *Leguminosae* are widely distributed around the world and show a considerable variation in size and habit. Soya beans, ground nuts and chickpeas are derived from tropical or sub-tropical plants in this family, whereas in temperate climates, peas, beans, clover, lupins and gorse provide familiar examples of leguminous plants. They provide an important source of human and animal foods and they can also enrich soil considerably. Soil enrichment is achieved as a result of the nitrogen fixation of their bacterial partners.

If a legume seed germinates in soil containing a population of rhizobia, both the legume and the bacteria interact to form root nodules on the growing plant. The seedling releases a variety of chemicals into the soil and these encourage the growth of the bacterial population. Among the most important of the growth stimulators is homoserine. The presence of rhizobia in the soil also causes structural changes in the growing legumes. The first visible effects of the bacteria are the branching and curling of root hairs (see Fig. 1.2). Bacteria nearby then multiply and invade the host in an **infection thread**. Bacteria have plant cell wall material deposited around them as they grow into the plant through the infection thread. This grows through the root hair cell and penetrates other root cells nearby, often with a considerable degree of branching of the thread. The root cells then proliferate to form a root nodule.

When they are free-living, rhizobia adopt a different shape from the bacteria found in root nodules. The free-living forms have a much more regular structure than do rhizobia found within root nodules, where they exist as irregular cells called **bacteroids**. Inside root nodules, bacteroids have increased membrane surface areas to aid metabolic exchanges between the plant and the bacterial cell. The cell wall material of bacteroids is also more permeable than that of free-living bacteria. This also facilitates metabolic exchange. Such constraints do not apply to rhizobia when they are not located within nodules.

Root nodules typically appear pink in colour because of production of a form of haemoglobin called **leghaemoglobin**. The enzyme responsible for nitrogen fixation is nitrogenase and it is easily damaged by the presence of oxygen. Leghaemoglobin acts as an oxygen buffer for the root nodule. It provides sufficient oxygen for the metabolic functions of the bacteroids, but it prevents the accumulation of free oxygen that could otherwise destroy the activity of nitrogenase.

There are over 100 plant species other than legumes that form root nodules containing nitrogen-fixing bacteria of the genus *Frankia*. These bacteria are slow growing and are difficult to cultivate in the laboratory. When found in nature, they form **actinorrhizas** – analogous to the mycorrhizal partnerships between the roots of vascular plants and fungi. Some of these plants are shrubs and trees, for example the Alder. Alders are currently being used in forestry to provide a natural supply of fixed nitrogen for the forest soil, enriching the environment for more economically valuable trees.

2.4   From what infections do plants suffer?

Microbial infections of plants have an enormous impact upon humans. They can have a devastating effect upon human crops, causing famine as well as economic loss. Furthermore, fungal infections of crops can cause the production of potent toxins that can directly affect the health of humans and animals that eat these crops. All groups of microbes may cause plant diseases but fungi are numerically the most important plant pathogens. The most destructive plant pathogens are bacteria. In contrast, horticulturists may even exploit virus infections of plants. The variegation of flower colour seen in tulips may result from a virus infection, and gardeners consider this a desirable feature for their specimens. This raises the question of when is a disease not a disease? The answer is presumably when it produces effects that please humans.

2.4.1   What plant diseases are caused by fungi?

Economically the most important plant infections are those caused by fungi. They have been recognised for centuries and can cause a range of diseases in crop plants including cereals, tubers and fruit plants. They are the bane of the gardener's existence. Fungi cause mildews, rusts, smuts, blights and scabs. Each of these may strike fear in the heart of any horticulturist, but the farmer is equally concerned about fungal infections of crops. They can spread rapidly and may disperse their infectious spores over large areas, bringing devastation of entire crops. The most extreme instance of the damage done by a plant fungal disease is the Irish potato famine.

Phytophthora infestans was the fungus responsible for the Irish potato famine in the middle of the nineteenth century. Over two million people died of starvation because of the repeated failures of potato crops between 1845 and 1849, since in nineteenth century Ireland, potatoes were the staple food for most of the peasant population. More than a million survivors emigrated to the USA as a result of the famine. The devastation can only be imagined, especially since Ireland had a population of just over eight million at the start of the famine. This fungus is still active today and can cause late potato blight. It is a disease of mature plants that starts with pale green spots appearing on the leaves of the plant. These turn very dark, becoming almost black as the plant dies. The fungus disperses spores into the soil ready to infect the next crop. Wind dispersal can scatter spores over a wide area. The infection may also persist through the winter in infected tubers. Although potato strains that are more resistant to infection have played some part in the control of this disease, these days the earlier harvesting of potatoes prevents the disease from developing. Other economically devastating crop infections caused by fungi have included the powdery mildew of the grape (*Erisyphe graminis*) that all but wiped out the French grape harvest when the fungus was imported from America into France. Similarly, Sri Lankan coffee plantations were devastated by *Hemileia vastatrix*, a fungus that attacks the roots of the coffee plant.

The mycelial nature of fungal pathogens is apparent in the mildews, where fine wefts of mycelia can be seen covering the infected tissues. There are Old Testament accounts of mildews, describing the pestilence they cause and comparing their effect to the devastation caused by locusts. The rusts get their name from the dark orange-red spots that appear on infected plant tissues. This coloration within the lesions is a consequence of the presence of a large number of spores; smuts are similarly caused by black spores. Besides infections that cause tissue destruction, some fungi produce chemicals that act as plant growth promoters. These can cause the development of plant tumours known as galls, although the most famous plant gall is caused by the bacterium *Agrobacterium tumefaciens*.

The nature of fungal attack is very variable. In some cases the pathogen attacks only selected tissues, causing a clearly defined infection. An example is brown rot of fruit caused by *Sclerotinia fructigena*. This infection is confined only to the fruit of an affected plant; the remainder of the plant is healthy. In contrast, the potato blight fungus *Phytophthora infestans* causes an infection that spreads through the entire plant. Different tissues of the same plant may be subject to infection caused by different species of fungus. An example is provided by the rusts of wheat plants. Leaf rusts are caused by *Puccinia recondita* whereas stem rusts are caused by the related *Puccinia graminis*.

Fungi can have a complex relationship with plants. Depending upon the prevailing conditions, they may live symbiotically in a mycorrhizal relationship or, when conditions change, the fungal partner may become an **opportunist pathogen**. Other fungi are **obligate parasites**. These fungi cannot be grown in artificial culture even on complex media and must rely on their host for their continued existence. It is very rare for obligate parasitic fungi to cause fatal diseases in their host plants.

Some fungi are primary pathogens; others are opportunists and only infect previously damaged tissues. *Botrytis cinerea* is an example of a fungus that is an opportunist pathogen. It causes strawberry grey mould but affects only fruit that has already sustained damage. Its normal habit is to live as a **saprophyte** on the surface of the plant, where it causes no apparent damage. Primary pathogens must enter undamaged plant tissues. This may be achieved through the roots in a manner similar to that used by fungi to establish benign mycorrhizal relationships. Another portal of entry for fungi is the stomata (singular: stoma): openings on the leaves that allow the diffusion of gasses into plant tissues. Pathogenic fungi often produce specialised branching structures known as **haustoria** when infecting plant cells. Many fungi are capable of infecting plant seeds and so the infection can be passed vertically from one generation to another. Pathogenic fungi may also be spread by insects. Dutch elm disease, caused by the fungus *Ceratocystis ulmii*, is spread by the bark beetle

Scolytus multistriatus. These insects penetrate the bark of trees and there lay larvae that burrow through the wood to form a maze of tunnels. Because of this behaviour, this insect is also known as the engraver beetle. Spores of *Ceratocystis ulmii* are carried by insects from one tree to another, thus spreading infection.

Dutch elm disease is so called because it was first described in elm trees (genus *Ulmus*) growing in Holland in 1919. The disease process begins by causing the leaves of an infected tree to turn yellow and then to drop. The infection is rapidly fatal as it spreads through the entire tree within a few weeks. The disease was first seen in the UK in the 1920s but it did not become a serious problem until the 1960s. In the late 1960s and early 1970s the disease became rampant in southern England and caused devastation of the elm population across most of the country. During the 1970s effective control of this disease was developed by injecting fungicide into trees that showed the first sign of disease. Badly affected trees were cut down and burned, as was any elm wood that may have been harbouring spores of the fungus. Today, strains of elm that are resistant to Dutch elm disease are being planted to replace those lost to the disease. It will, however, be many years before these trees reach maturity.

There is a devastating disease of rice seedlings called *bakanae* disease, meaning foolish seedling disease. Affected plants seem to bolt and very rapidly outgrow their unaffected neighbours. This pattern of growth leads to a tall, straggly plant that is too weak to support itself. The cause of this disease was discovered in Japan in 1926. Growth is caused by infection with a fungus known as *Gibberella fujikuroi* with an asexual form *Fusarium moniliforme*. It produces excess amounts of a plant hormone of the gibberellin family. These hormones are active at a concentration around one part per million and are effective on a wide variety of plants other than rice seedlings. Because Western scientists did not read the scientific literature published in Japan, knowledge of gibberellins was not widely disseminated and gibberellins were 're-discovered' in the early 1960s by Western scientists. It was only later that the older Japanese work was discovered. In this case, the story of fungal infection of plants may have a happier ending than many. The gibberellins are commercially exploitable as plant growth promoters. Life is, however, never simple. One of the first commercial uses was to stimulate the growth of hemp plants to make longer fibres for sacking, rope manufacture, etc. It should be remembered that the hemp plant is *Cannabis sativa*. Perhaps a safer economic use is in the elongation of edible celery stalks.

Much effort is expended in trying to control fungal plant pathogens, and a variety of measures have been developed. The importance of such controls is reflected in the strict legal controls governing the import of potentially infectious plant material imposed by many countries. Once a fungal infection has become established, then radical treatment is the best option. Infected material must be gathered and destroyed, preferably by burning. This is the only way to ensure destruction of the spores that can otherwise spread to cause infection in neighbouring plants.

Crop rotation is an important measure used to control fungal infection of plants. By leaving ground free from vulnerable plants for variable periods, the numbers of spores of pathogenic fungi present in soils will diminish and, as they disappear, the chance of infection also falls. Similarly, the use of pathogen-free seeds or tubers helps to reduce the chance of an infection developing. Probably the greatest success in the control of fungal pathogens has come from the development of a range of chemical fungicides. The long-term effect of fungicides on food crops has not yet been evaluated and there must be a degree of concern about the intensive use of such chemicals. Biological controls would appear to provide an attractive alternative to the use of chemicals. It has, however, proved remarkably difficult to develop effective biological controls. A limited success has been achieved with some plants where fungal infection has been prevented experimentally by increasing the number of saprophytes on the surface of the plant.

It is not just vital plant tissue that can be attacked by fungi. Even though it is a highly durable natural product, wood must also be broken down over a period. If this were not so, the Earth would be entirely covered in accumulated dead wood and vital elements would be gradually diminished as they remained locked in the dead trees. Indeed, the durability of wood can be demonstrated by its spectacular persistence, and although ancient dead wood is not apparent on the surface of the Earth, we do have abundant supplies of coal. Most coal deposits are wood and other plant material that has been preserved under anaerobic conditions and compressed to appear more like a mineral than plant matter. Anyone who has tried to maintain a garden fence will know, however, that wood can easily rot and if the conditions are appropriate can decay at an alarming rate. Fence posts seem particularly vulnerable to wood rot. This is because they spend their time standing in soil. The soil microflora treat fence posts as they would any wood: a source of a long, slow banquet, gradually releasing its nutrient for re-cycling.

Wood is derived from the vascular tissue of plants and its principal components include celluloses, hemicelluloses and lignin. These are complex polymers of sugars and alcohols that are highly resistant to biodeterioration. The precise composition of wood varies from species to species but does broadly reflect the division of timber into hard- and softwoods. Hardwoods are derived from deciduous trees; softwoods come from conifers. These terms may, however, be misleading. Balsa wood, beloved of the makers of model aeroplanes, is *technically* a hardwood.

Timbers differ greatly in their ability to withstand fungal attack. Because of their durability, tropical hardwoods are in great demand and this has led to deforestation of large areas in central Asia and tropical Africa. One of the most resistant of timbers is oak (*Quercus robur*). It also has enormous mechanical strength. This was the wood chosen to replace the roof beams in the reconstruction of the south transept of York Minster, destroyed by fire in 1984. It is also the wood used to build ships of the British Navy in the days of sail. Heartwood, taken from the centre of the tree, is more durable than the outer timber. Is this why the Navy adopted 'Hearts of Oak' as a favourite march tune?

A few specialised bacteria, evolved to live in the stomachs of ruminant animals or in the guts of wood-boring insects for example, have acquired the knack of digesting wood. The most common organisms to attack woods, however, are species of basidiomycete fungi. These are the fungi with fruiting structures that include toadstools, mushrooms and puffballs. Bracket fungi may even attack live wood and can be seen growing out of the trunks of living trees.

Wood rot is of enormous importance since wood is a major structural component of many homes. It is because poor-quality timbers were used for window frames in houses built in the 1960s and 1970s that the uPVC double-glazing industry became rampant across the UK during the 1980s. Over time the timber rots away. One of the most dreaded wood rots, which affects older properties as well as new houses, is dry rot, caused by the fungus *Merulius lacrymans*. This fungus belongs to the family that cause 'brown rot' of wood. They digest the cellulose and hemicellulose component of timber but leave the brown lignin polymers largely unaffected. Removal of the cellulose and hemicellulose from wood also removes its mechanical strength. Dry rot fungi produce flat fruiting structures that cover the surface of the infected timber, but the area of infected wood extends far beyond the fruiting structure. Surrounding that area, the wood becomes divided into small brown cuboid sections as the fungus digests away the cellulose in the wood. Radical treatment is required to treat dry rot. The best course of action is to remove the affected wood entirely and to replace it with new, well-seasoned material. The seasoning of wood involves the controlled drying of timbers. This reduces the water content, making the timber less vulnerable to infection. Replacement timbers may also be treated with chemical preservatives to help to prevent infection. A number of effective treatments are available. This is important since the mycelia of the dry rot fungus can penetrate brickwork and other apparently solid barriers.

Of greater ecological importance than the brown rot fungi are the white rot fungi. These fungi can digest not just cellulose but also eat away the lignin present in woods. It is the white rot fungi that are responsible for the decomposition of dead wood on forest floors. Some white rot fungi do not digest lignin completely but are classified as causing white rot because they cause bleaching of the lignin remaining in the wood. Others attack lignin in preference to cellulose and only digest the cellulose component of timber when the supply of lignin becomes exhausted. Yet other white rot fungi attack cellulose and lignin simultaneously. In pure culture, white rot fungi take many months or even years to cause the experimental destruction of woods. Wood decay can be considerably more rapid when a mixed population is used. White rot can only occur in the presence of oxygen, and white rot fungi also require a supply of glucose to digest lignin. The cooperation of more than one fungus growing on timber increases the supply of glucose derived from the breakdown of cellulose in the wood.

2.4.2   What plant diseases are caused by bacteria?

Although not numerically the most important of plant pathogens, bacteria cause some of the most destructive of plant diseases. The majority of plant infections caused by bacteria result in the collapse of tissues. Galls, however, represent the undifferentiated growth of plant tissues and provide an alternative type of pathology caused by bacteria. Besides galls, bacteria cause leaf spots and leaf stripes, wilts, cankers, blight and rots. Bacterial infections may cause retardation of fruit ripening and the deformation of fruits. Most plant pathogens are Gram-negative bacilli, although species of the Gram-positive genus *Corynebacterium* also act as plant pathogens. Many bacteria that cause plant diseases also grow as soil saprophytes.

The growth of crown galls is initiated by *Agrobacterium tumefaciens* strains carrying a Ti plasmid, so called because of its tumour-inducing properties. The Ti plasmid is transferred from the bacterium into plant tissue. Once the Ti plasmid enters the plant cell it becomes incorporated into the plant genome. It then stimulates the growth of undifferentiated plant tissue, forming the tumour. The undifferentiated cell growth causes deformation of neighbouring tissues, cutting off the flow of water and nutrients to other parts of the plant, ultimately causing its death. After it has transferred the Ti plasmid, *Agrobacterium tumefaciens* plays no further part in the pathogenic process. The Ti plasmid has been exploited to introduce foreign DNA into genetically engineered plants. This DNA encodes 'desirable' properties that enhance the economic importance of the plant. Such plants are also known as transgenic plants because they contain foreign DNA. Crown gall disease can now be controlled biologically by applying *Agrobacterium radiobacter* strain 84. This works by out-competing the pathogenic *Agrobacterium tumefaciens*.

Bacterial spots and stripes are most frequently caused by pseudomonads such as *Pseudomonas syringae* and *Xanthomonas campestris*. Along with leaf spots and stripes, bacterial infection may result in the appearance of spots on flowers and on fruit.

Slime-producing bacteria that grow in the xylem of plants produce wilts. This interferes with the transport of water and solutes through the plants. Consequently, the infected plant wilts and eventually dies as nutrient distribution is disrupted. Species of the genus *Erwinia, Pseudomonas solanacearum* and *Corynebacterium insidiosum* are among the bacterial species that cause wilts.

Cankers are areas of **necrosis**. Like wilts, cankers start in the xylem but they extend into the surrounding tissues. Bacteria that produce cankers also cause the damaged tissues to exude gums in the process of **gummosis**. Species that cause canker formation and gummosis include *Pseudomonas syringae, Xanthomonas campestris* and *Corynebacterium michiganense*.

Erwinia species are opportunistic plant pathogens. They cause the most destructive soft-rots of vegetables. They produce a host of exoenzymes, particularly pectinases, that break down the integrity of the plant cell, leading to loss of turgor and the destruction of the infected tissue, which becomes slimy. They are also responsible for the fire blight that can kill apple and pear trees within a single growing season. This is caused by the bacteria infecting the bark, where they produce exoenzymes that damage and block the plant's vascular system, which is essential for water and nutrient transport.

Wind and water can help to spread bacterial plant pathogens, as can the movement of infected soil. Horticultural and agricultural implements may also help to spread bacterial plant diseases. Arthropods may spread infections by *Spiroplasma* species and these cause stunting in many types of plant.

2.4.3 What plant diseases are caused by viruses?

Almost every commercial plant is affected by at least one virus infection. These can severely damage the commercial viability of the crop and in many cases also threaten the life of the plant. It has been estimated that plant viruses cause economic damage in excess of $70 billion annually. Worldwide, more than 50 million citrus trees have recently been destroyed by the activity of a single virus; another has destroyed 200 million cocoa trees in Western Africa. It was the economic importance of plant virus infections such as tobacco mosaic disease that provided the impetus for much of the early virus research. This was facilitated because, compared with animal cells, plant tissue cultures are much easier to initiate and maintain.

Viruses are **obligate intracellular parasites**. They rely entirely on their host to maintain their lifecycles. It is for this reason that perennial plants that grow year after year are more vulnerable to virus infections than are annuals. Plant viruses are diverse and belong to different families. Over 1000 individual plant viruses have now been recognised. Typically plant viruses are named by a combination of the name of the host and the type of disease it produces. For example, potato yellow dwarf virus is a virus that infects potatoes causing the plants to turn yellow. As with animal viruses, plant viruses have either an RNA or a DNA genome. The genome, whether RNA or DNA may either be single or double stranded. This is unusual because no known human DNA virus contains single-stranded DNA. Plant viruses may be enclosed in a lipid envelope or they may have a naked protein coat enclosing the genome.

Virus infection of plants can have many different effects on the infected host. A common feature of virus infection is loss of pigmentation, causing spots, stripes or mosaics. In some cases, this may cause the death of plant tissue, forming areas of brown necrosis. If phloem tissue is targeted for virus infection, this results in stunting of the plant since the nutrient transport system is disrupted by the virus infection. Virus infection may lead to uncontrolled growth. This can lead to misshapen fruit, as with cucumber mosaic virus, or the formation of plant tumours. The virus load in a plant can become very high. It has been estimated that 10% of the dry weight of a heavily infected tobacco plant affected by tobacco mosaic virus is accounted for by virus particles.

Because of the problems of getting viruses together with a sedentary host, many plant viruses are highly infectious and can maintain infectivity for a long period. Smokers who grow *Nicotiana* plants can pass on tobacco mosaic disease from virus particles in their cigarettes. The tobacco mosaic virus can also remain infectious for very long periods. Estimates of infectivity have been upwards of 50 years.

The tobacco mosaic virus is unusual in that it relies upon entry into damaged plant tissue. Many viruses rely upon insects to act as vectors to transfer them from one host to another. The viruses may simply attach to the mouthparts as the insect feeds on an infected plant and are then transferred to the next vulnerable plant. Alternatively, the insect may ingest the plant virus along with its feed and the virus may remain within the body of the insect. Although plant viruses cannot replicate in the vector, the insect may remain a **carrier** for the rest of its life. Since insects spread so many plant viruses, control of the insect vector is the method most commonly used to prevent spread of plant virus infection. Many arthropods can act as vectors for plant viruses. These include aphids, leaf hoppers, mealybugs, whiteflies, thrips and mites. Rarely, infected pollen grains may carry virus particles from one plant to another. Worms are another occasional vector of plant viruses, as is the human practice of grafting, where plant tissue from one source is transplanted to the stock of another strain. In agriculture and horticulture, plant viruses may be spread by the use of contaminated implements used for plant propagation. Plant surfaces provide a protective barrier against infection and if a plant virus is to infect a new host, this barrier must be breached. This is why so many plant viruses require the intervention of a vector. Others, such as the clover wound tumour virus, rely upon accidentally damaged tissue if they are to cause an infection.

Some viruses have aesthetically pleasing qualities. Horticulturists have long admired 'tulip break'. This is the striped colour variegation seen in the flowers of prized specimens. This feature can be passed from one generation to the next and was at one time thought possibly to be a genetic trait. We now know that tulip break is caused by a virus. The virus can enter the bulb and thus can be propagated when infected bulbs split.

Plants are also affected by **viroids**. These are small RNA molecules that are not associated with any protein coat and that can cause infection of plants. The RNA molecule is much smaller than the genome of typical RNA viruses, being only about 200 kilobases. Viroids have a complex secondary and tertiary structure that makes them resistant to the action of nucleases, which would otherwise digest the RNA. The first viroid to be described was reported in 1962 and causes potato spindle tuber disease. There has been continued speculation that viroids could infect animals as well as plants. It was proposed that the transmissible spongiform encephalopathies result from viroid infection. The weight of evidence would now suggest that infectious proteins cause these diseases and at present no animal infection is attributed to a viroid.

Many of the early virology studies exploited tobacco mosaic virus. This virus causes a mosaic pattern to appear on the leaves of infected plants. The infectious disease was first described as long ago as 1898, and in 1935 the virus particle, then described as a 'globular protein', was first crystallised. This was the first virus to be examined in the electron microscope.

Once a plant has been infected with a virus it is impossible to eradicate the infection. Plant virus infections have traditionally been controlled by destroying infected plants and by initiating breeding programmes to develop resistant lines. Plants do not have an immune response such as that which protects us from infection. More recently, molecular biologists have been applying the techniques of genetic engineering to the problem of plant virus infections. As with the early virus studies, much of this research has centred upon tobacco mosaic virus. Transgenic tobacco plants have been produced that show marked resistance to infection. The mechanism by which this resistance is conferred is not at all clear. This opens up a fascinating area for future research in the field of plant pathology.

2.5 How are microbes used to control agricultural pests?

For most of the twentieth century, we have been developing an enormous range of chemicals to control agricultural pests and to improve crop yields. Despite the initial enthusiasm for such products, we slowly became aware that their use brought considerable risks. At one time the insecticide DDT (dichlorodiphenyltrichloroethane) was hailed as a great innovation that would control a whole host of diseases spread by insects. We now know that DDT is highly toxic to other life-forms and that its effects are cumulative. Because of this, DDT is now banned from worldwide use other than in exceptional circumstances. Exposure to organophosphates such as DDT can cause an illness characterised by muscle pain and chronic fatigue. The risks of the long-term use of other chemicals have yet to be properly evaluated. It has been argued that it may be better to use biological controls rather than rely on artificial chemicals. There have been some notable successes but the implementation of genetic engineering technology should be approached with due caution. Unless carefully regulated and monitored, the use of genetically engineered organisms may be creating biological problems greater than those caused by exploitation of artificial chemicals. However, the introduction of an insecticide gene into a plant significantly reduces the risk that the plant will suffer fungal infection. In turn, this will reduce the risk of human exposure to mycotoxins, for example.

When species are transported from their natural habitat to lands where they have no natural predators, the effects can be devastating. In the late 1880s the gypsy moth (*Lymantria dispar*) was imported from France into the USA. The intention was to use it for silk production, but the venture was not a success. The moth escaped into the woodlands where it has wreaked havoc for a century. It is a serious pest in orchards. In damp weather, however, this moth succumbs to attack from a fungus, *Entomophaga maimaigma*. This fungus has been used to control the gypsy moth but it has not been universally success-ful. The usefulness of fungal insecticides is limited by our inability to control the climatic conditions necessary to cause a reasonable level of fungal infec-tion in the pest population.

A similar problem arose with the introduction of European rabbits (*Oryctolagus cuniculus*) into Australia. By the 1950s the rabbit pest problem had reached epidemic proportion. They had colonised vast areas of the con-tinent and grazed the grass to destruction. This threatened ruin to the Australia sheep farmers whose animals required adequate pastures to feed. In an attempt to control the problem, myxomatosis virus was introduced into Australia to try to control the rabbit problem. The virus is *endemic* in South America where it causes a mild infection in rabbits of the genus *Sylvilagus*. This scheme met with partial success. In dry areas, the virus did not spread because it is transmitted by a mosquito vector that requires ponds in which to breed. There were additional problems. In the early days of the experiment, the majority of rabbits succumbed to infection. With time, however, a rabbit population that was resistant to myxomatosis was selected. Furthermore, when the virus was examined, it was found to have undergone mutation to select less virulent strains. Another problem was that the infection spread to Europe, causing problems for the native rabbit population.

Baculoviruses can infect a range of insects and have been used to control infestations of corn ear worm, cotton boll worm, soybean pod-feeding cater-pillar and the sawflies that affect spruce and pine woods. The baculovirus has the advantage that once it has been used to inoculate pests in an area needing control then the infection will become self-sustaining. This will continue until the pool of pests that can act as potential hosts is too small to sustain the con-tinued replication of the virus. It also has the advantage of a relatively limited insect host range that spares many of the insects that do no damage to crops; hence it is a specific control agent. One problem is that infected insects may, in the short term, increase their food intake considerably. This may cause more damage to the crop than the original infestation. To overcome this, insect-specific toxins have been genetically engineered into baculoviruses to increase the efficacy of killing and to reduce the damage done to crops by infected insects. Indeed, because of the ease with which they accept and express cloned DNA, baculoviruses are becoming an increasingly important tool in the study of eukaryotic genes.

Bacillus thuringiensis is a Gram-positive spore-forming bacillus. During

sporulation this bacterium forms a protein crystal around the spore, sometimes called the parasporal body or parasporal crystal. This acts as a toxin precursor. When insects swallow the bacteria the parasporal body is digested to
release an active insecticidal toxin. Different strains of *Bacillus thuringiensis*
produce variant toxins, each with a different target insect. The specificity of
the *Bacillus thuringiensis* toxin depends upon the presence in the insect gut of
an enzyme that can release the active toxin from the parasporal body precursor. *Bacillus thuringiensis* is successfully used to control populations of insect
pests such as the alfalfa caterpillar, the cabbage worm and the gypsy moth.
Attempts have been made to select a strain capable of controlling the
Anopheles mosquito, vector of malaria, but these efforts have met with only
limited success. Other members of the genus *Bacillus* have insecticidal properties. These include *Bacillus popilliae* and *Bacillus lentimorbus*. Both species infect
the larvae of Japanese beetles. *Bacillus sphaericus* is currently being investigated
for its potential in the control of the malaria mosquito.

Damage to crops comes not only from a biological source. Natural phenomena can devastate crops. In the UK, gardeners live for several months
with the threat of frosts destroying the products of their labours. So important is this problem that the Meteorological Office regularly issue frost warn-
ings on their broadcast weather forecasts. The first frost of the late autumn
will destroy all nasturtiums completely. As water freezes it expands, causing
weaker plant cells to burst. The problem for many crop plants is made worse
because the bacteria that live as saprophytes on their leaves act as foci for ice
crystal formation. A protein on the surface of 'ice nucleating' bacteria that
acts as the primary focus for ice formation has been identified. Genetic engineers have now produced 'ice minus' mutants of the plant saprophyte
Pseudomonas syringae, in which the protein that acts as a focus for ice formation
is not made. These bacteria can be applied to vulnerable crop plants, where
they can reduce drastically the number of ice-nucleating bacteria and in so
doing can prevent frost damage.

The Ti plasmid of *Agrobacterium tumefaciens* is increasingly being exploited
to introduce genes that encode 'desirable' properties into crop plants. It is also
used as a vector in antisense technology. According to the central dogma of
molecular biology, double-stranded DNA molecules are transcribed to
produce single-stranded messenger RNA (mRNA). This is then translated
into a protein product. In antisense technology, DNA that encodes a complementary mRNA copy is engineered and introduced into an appropriate host
cell. When the first gene is transcribed, so too is the antisense gene. The two
RNA molecules will then pair together and the natural gene product can no
longer be formed. Antisense technology and the Ti plasmid have been

successfully used to delay overripening of tomatoes, increasing significantly the shelf life of the fruit.

Maize is an important crop used to feed both humans and domestic animals. It is also increasingly being employed in the production of **silage**, used as an over-winter feed for farm animals. There are a number of diseases that affect maize. To help to improve disease resistance, transgenic maize has been engineered to withstand disease and to resist herbicides. This enables maize fields to be weeded easily with applications of a weed killer to which the transgenic crop is resistant. One vector used to produce the transgenic maize is based upon the bacterial pUC cloning vectors; these carry an antibiotic-resistance gene that confers resistance to ampicillin and other penicillins. This gene is not expressed in the plant tissue. There is, however, a small but finite risk that it may transfer from the transgenic plant into the microflora of an animal that is fed on this crop. Even more concerning is the observation that the gene coding for ampicillin resistance can undergo point mutations. These extend the antimicrobial activity of the gene product to include the newer cephalosporins, which are used for the treatment of life-threatening infections. Were these to get into the gut microflora, a considerable threat would be posed. Although the USA has licensed this crop for widespread use, its exploitation in the UK is currently still closely regulated. Because of such risks, caution should be exercised when using such products.

3

The microbiology of drinking water

Water is essential for the maintenance of all life on Earth. It also acts as the vector for many diseases caused by bacteria, viruses, protozoa and worms. For water to be regarded as **potable**, i.e. of a quality fit and safe for drinking, it must be free from such pathogens. Furthermore, it must not contain any other noxious substances such as chemical hazards including pesticides, insecticides or herbicides, artificial fertilisers or heavy metal ions. Potable water should not have an unpleasant odour or taste.

3.1 What are water-borne diseases?

Among the bacterial infections that are spread by water are cholera, the enteric fevers and dysentery. Hepatitis A and poliovirus cause infections after drinking contaminated water. Amoebic dysentery is caused by the protozoan *Entamoeba histolytica* and is spread either by drinking contaminated water or by eating food such as fresh fruit, salad or raw vegetables that have been washed in contaminated water. Other protozoal diseases such as those caused by *Giardia intestinalis* (*Giardia lamblia*), *Balantidium coli* and *Cryptosporidium* species are spread in a similar fashion. Schistosomiasis, also known as bilharzia, is a water-borne infestation caused by worms of the genus *Schistosoma*.

3.1.1 Cholera

Cholera is a disease that has been known since ancient times. It was confined to the Indian sub-continent but between 1817 and 1923 there were six *pandem-*

ics in which cholera spread from its original home across the world. It is now also endemic in South America as well as in Asia. Evidence of these great pandemics survives today as cholera graveyards can still be seen. One such graveyard exists outside the city walls at York, just opposite the railway station. One hundred and eighty-five bodies were interred there between 3 June and 22 October 1832. The city wall runs alongside the graveyard, with the graves lying outside the ancient boundary. This continued an ancient practice of burying the victims of infectious diseases beyond city boundaries.

The connection between cholera and water was made by the pioneering epidemiological studies of the anaesthetist John Snow in the early part of the nineteenth century in London. He observed that cholera cases were linked with people who obtained their drinking water from the Broad Street pump. To confirm his suspicions, he removed the handle from the pump, thereby preventing access to the contaminated water supply. This action caused a dramatic reduction in the number of cholera victims, confirming Snow's hypothesis and saving many lives. In the Indian sub-continent, cholera is most common along the pilgrim routes and is found particularly along the Sacred Rivers such as the Ganges.

The pandemics of the past were caused by the 'classical' serovar of *Vibrio cholerae*. This serovar causes the most severe form of the disease. Between 1960 and 1971, a seventh pandemic occurred. It was caused by the El Tor strain of *Vibrio cholerae*. This was named after the village in Sinai where it was first isolated. The El Tor strain causes a milder form of cholera, associated with a lower rate of mortality than the disease caused by the classical strain of the bacterium. The El Tor strain has now replaced the classical strain as the endemic strain around the world, except in parts of the Indian sub-continent.

Vibrio cholerae is a pathogen found only in association with humans. The incubation period for cholera is between one and four days and the infectious dose is very high. It has been estimated that it requires a dose of 10^{10} bacteria to initiate an infection. Symptoms of cholera include fever, vomiting and profuse watery **diarrhoea**. As the disease progresses, the patient's stools take on the appearance of water in which rice has been boiled, hence the description of 'rice-water stools'.

Healthy adults typically lose about 2.5 litres of fluid daily: in cholera victims it can amount to between 10 and 15 litres per day. This is because of the action of the cholera toxin, known as **choleragen**. This raises the levels of intracellular cyclic AMP in the epithelial cells lining the gut. This reverses the sodium pump, causing a disturbance of the electrolyte balance of the body. High concentrations of salts build up in the lumen of the gut and this causes fluid loss from the body as a result of osmosis. This results in circulatory collapse and death. Untreated, cholera has a mortality rate of up to 50%. This can be markedly reduced by fluid-replacement regimens and still further reduced by antibiotic therapy. The drug of choice is tetracycline. Remarkably, histological examination of the gut from fatal cases of cholera shows a perfectly normal cellular appearance. Cholera toxin causes no apparent damage at the tissue or the cellular level. Rather, its effects are to be found at the molecular level. Because of its effects upon the electrolyte balance of the body, however, its consequences can be devastating.

3.1.2   Enteric fever

Enteric fever is the collective term given to the invasive infections caused by *Salmonella typhi*, the cause of typhoid fever, and by the strains of *Salmonella paratyphi* that cause paratyphoid fever. Like *Vibrio cholerae, Salmonella typhi* is a pathogen that only has humans as its natural host. The infectious dose is generally about 10^5–10^6 bacteria. This dose may be lower if the patient has been taking antibiotics or antacids when exposed to the infectious agent.

The incubation period for typhoid fever is approximately two weeks. During the incubation period, the bacterium colonises the small intestine. From there it enters the lymphoid tissue and thence it reaches the bloodstream. Typhoid represents an invasive infection, with bacteria surviving within phagocytic white blood cells. Laboratory diagnosis most often relies on isolating *Salmonella typhi* from blood cultures. No enterotoxin has been identified.

Typhoid infection tends to localise in the spleen, liver and gall bladder. When the bacterium enters the bloodstream it causes **septicaemia** (blood poisoning). The first clinical symptom is fever, coinciding with the onset of septicaemia. The symptoms of typhoid are largely caused by **endotoxin**: the lipopolysaccharide associated with the bacterial outer membrane. During the early stages of the disease, patients suffer from a dry cough, headache and abdominal pain. Constipation is a very common symptom during the early stages of enteric fever. As the disease progresses, areas of inflammation occur in localised lymphoid tissues known as Peyer's patches. After about two weeks, patients develop diarrhoea. At about this time, 'rose spots' appear on the patient's abdomen as a result of **haemorrhage** under the skin. This rash resembles the rash of typhus, caused by *Rickettsia prowazekii*. This is how the name typhoid was derived: one of the symptoms resembles typhus. The two diseases are otherwise unrelated. Later complications of enteric fever include haemorrhage, intestinal perforation, **osteomyelitis** and **meningitis**. Untreated cases of typhoid are associated with a mortality of about 10%.

Bacteria that carry a Vi antigen cause the most severe disease. 'Vi' is short for virulence, so-called since strains with this antigen cause a more aggressive disease than those without it. The antigen is now identified with capsular material that affords protection from our immune defences. Most isolates associated with human disease have a capsule, at least upon primary isolation. The capsular material may be lost upon ensuing sub-culture. *Salmonella typhi* is an intracellular parasite and its capsule protects the bacterium from our natural defences as the organism hides inside macrophage cells. About 5% of typhoid patients become asymptomatic carriers, often harbouring the bacterium in the gall bladder. A small number of carriers excrete *Salmonella typhi* in their urine.

Water is the principal vector for typhoid fever but this need not necessarily be as drinking water. Ice used in drinks and the water used to wash salads, fruit and vegetables have been found to cause typhoid. The symptoms of paratyphoid fever are similar to those of typhoid but *Salmonella paratyphi* is associated with a milder disease and causes fewer deaths.

3.1.3 Bacilliary dysentery

Bacilliary dysentery is caused by bacteria in the genus *Shigella*. Clinically it manifests itself as a 'bloody flux': diarrhoea typified by the presence of blood, pus and mucus. There is also a severe abdominal pain associated with the condition. The most severe form of the disease is caused by *Shigella dysenteriae* and this species is particularly prevalent in the Third World, where a high mortality rate is found among untreated patients. In the UK, most cases are caused by *Shigella sonnei*. This causes a milder dysentery. *Shigella sonnei* is unusual in that it can ferment **lactose** but only after a prolonged incubation period. While the colonies look like non-lactose fermenters after overnight incubation, by the end of a second day they appear as lactose fermenters. Consequently, they are described as 'late lactose fermenters'. Two other species, *Shigella flexneri* and *Shigella boydi*, cause a disease of intermediate severity: more serious than that caused by *Shigella sonnei* but less severe than *Shigella dysenteriae* dysentery. In the case of all shigellas, infection causes invasion of tissues in the gut mucosa. Inflammation is, however, confined to the sub-mucosa and it does not spread beyond this region. *Shigella dysenteriae* elaborates a toxin known as the **Shiga toxin.** This has neurotoxic properties and is thought to be responsible for inducing coma in fatal cases of dysentery.

In the UK, cases of *Shigella sonnei* dysentery are most frequently seen in nurseries and play schools, although they also often occur in psychogeriatric institutions. The disease is spread because of the poor personal hygiene observed by small children and psychogeriatric patients. This problem is exacerbated if the staff in such institutions are overworked. Infection under these circumstances probably spreads as a direct consequence of direct faecal–oral contact, often by way of dirty fingers. Small children love to suck their fingers but may not be very careful about washing their hands after going to the toilet. Shigellas can readily pass through toilet paper. They can get through hard 'sanitary' paper as well as through soft papers. The infectious dose of shigellas required to initiate dysentery is very small, about 100–1000 bacteria. This facilitates the spread of infection in situations such as nurseries.

3.1.4   Water-borne campylobacter infections

Water-borne campylobacter infections are being increasingly recognised. Many wild birds act as a natural reservoir for these bacteria, as do domestic animals and birds. Faecal material easily contaminates lakes and rivers that feed reservoirs used for drinking water. Failure of the decontamination system for domestic water supplies and direct faecal contamination of domestic water storage tanks has been responsible for incidents of human campylobacter infections. This problem is compounded by the fact that in natural waters campylobacters undergo a transition to a viable but non-cultivable state in which they may persist for considerable periods. In experimental animals, these non-cultivable campylobacters have been shown to cause infection. Furthermore, when symptoms of infection are manifest, the non-cultivable forms have been shown to revert to a state in which they can grow on artificial medium once more.

For many years, histopathologists suggested that duodenal and gastric ulcers may be caused by an infection rather than because of 'stress' or the other miscellaneous factors that were cited as possible causes. They made this claim because upon microscopic examination of biopsy material they saw what they thought were spiral bacteria. Microbiologists could not isolate anything from the material using conventional culture and, because of the extreme acid conditions in the stomach, they were very scathing about a microbial cause for ulcers. In the mid-1980s, however, and exploiting the technology used to grow campylobacters, 'campylobacter-like organisms' were isolated from ulcer biopsies. These bacteria can survive in the stomach because they produce a powerful urease that can split urea to release ammonia. This raises the pH in their local micro-environment, protecting the bacterium from acid attack.

Taxonomically, these bacteria have now been put into a separate genus and are called *Helicobacter pylori*. Previously they had been known as *Campylobacter pylori* and before that as *Campylobacter pyloridis*. They have been the subject of intense research and are now known to cause stomach ulcers. Definitive proof came from a self-inoculation experiment. As a consequence of such studies, the management of ulcers has now radically altered, with an emphasis on elimination of microbes from the ulcer rather than suppression of the symptoms they cause. As well as being the cause of stomach ulcers, there is now speculation that *Helicobacter pylori* may have a role in other diseases. These include heart disease and certain cancers, although the evidence is not as solid as that for their role as a cause of ulcers. Epidemiological studies carried out in Peru show that, in Third World conditions at least, this bacterium can be transmitted through contaminated water. As with the campylobacters, *Helicobacter pylori* can enter a resting, non-cultivable state, making its detection and recovery difficult.

3.1.5 Water-borne virus infections

Hepatitis A virus, an RNA virus of the picornavirus family, causes acute infective hepatitis, resulting in inflammation of the liver. This virus is very different from the other common hepatitis virus, the hepatitis B virus, which is a DNA virus, and the other hepatitis viruses (see Section 7.7.4). Although very different in structure and mode of spread, these two viruses cause diseases that are almost impossible to differentiate clinically. Infective hepatitis, caused by the hepatitis A virus, is also known as acute epidemic hepatitis. This reflects the sudden onset of symptoms in this generally self-limiting disease that frequently occurs in epidemics.

In the early stages of the disease, patients complain of fatigue, fever, diarrhoea, **malaise** and **anorexia**. In about two thirds of adult patients this so-called **pre-icteric** phase is followed by a period of jaundice that lasts for one to three weeks. Patients generally feel better once the jaundice develops. In contrast to adults, only about one in twelve children infected with the hepatitis A virus shows clinical evidence of jaundice. The disease is most often seen in children and young adults. It is more common in institutions where large numbers of people live close to one another, such as in military barracks or in boarding school dormitories. The infection spreads by the faecal–oral route and water is a common vector of hepatitis A infection. The virus first replicates in the epithelium of the gut. From there virus particles enter the bloodstream. They are then transported to the liver where they replicate, causing inflammation. Large numbers of virus particles are shed into the bile duct, where they regain access to the gut. The faeces of patients with hepatitis A are highly infectious.

Poliomyelitis is caused by the poliovirus, one of the RNA viruses that belong to the enterovirus family. The virus can replicate in the oropharynx or in the small intestine. Virus particles may be shed in droplets found in the nasal and throat discharges. They can, therefore, be transmitted by the respiratory route. Infected faeces are, however, also a significant source of human poliovirus infection and faecally contaminated water is an important vector for the disease. Other enteroviruses, notably coxsackieviruses and echoviruses can be transmitted in the same way as poliovirus. These viruses are important causes of heart infections and virus meningitis, respectively.

After poliovirus infects the mucosa of the oropharynx or small intestine, it replicates and spreads to local lymphoid tissue, particularly tonsils in the throat and Peyer's patches and lymph nodes in the abdomen. From there the virus enters the bloodstream to cause a transient **viraemia**. Even people vaccinated with the live, **attenuated** Sabin vaccine experience a viraemia. During the initial phase of the disease, patients suffer fever, sore throat, anorexia and headache. In most cases, the disease does not develop further than this. These cases are sometimes referred to as **abortive poliomyelitis**. In about 5% of patients, disease progresses to **aseptic meningitis,** lasting from two to ten days. Very few patients, about 0.1%, develop paralytic poliomyelitis. Virus particles spread along nerve fibres to the central nervous system, primarily attacking motor neurones. The obvious muscle wasting that is evident in patients with paralytic poliomyelitis does not result from the direct damage of muscle cells by the poliovirus. Rather, it is a secondary consequence of the damage caused to the motor neurones.

In addition to the viruses discussed above, there are a number of other water-borne viruses, e.g. rotaviruses caliciviruses, certain adenoviruses and the Norwalk agent. These all cause vomiting and diarrhoea. Some of these viruses are difficult to study because they do not grow in tissue cultures. These are discussed in Section 7.7.3.

3.1.6   Water-borne protozoal diseases

The cause of amoebic dysentery is *Entamoeba histolytica*. A common source of this condition is the consumption of contaminated water. As with other water-borne infections, this may not necessarily be as a result of drinking dirty water. In addition, certain homosexual practices have also been associated with the transmission of amoebic dysentery. There are many people worldwide who carry *Entamoeba histolytica* asymptomatically and it has not yet been established whether all strains have the capacity to cause disease. In patients who do develop symptoms, the disease has a more insidious onset than bacilliary dysentery. The condition is characterised by abdominal pain and cramps, with diarrhoea containing blood, pus and mucus. A milder form of the disease sees alternating phases of diarrhoea and constipation. The amoeba invades the mucosa of the intestine, causing multiple **abscess** formation. The presence of pus in the stools probably results from the secondary bacterial infection of sites already infected. Rare complications of *Entamoeba histolytica* infections include liver abscesses and amoebic pneumonia.

Giardia intestinalis (Giardia lamblia) causes infection of the upper portion of the lower intestinal tract. This protozoan cause chronic malabsorption and so patients pass loose, foul smelling, fatty stools. Giardia infections are characterised by diarrhoea, abdominal cramps, weight loss and anaemia. In the UK the illness is seen especially in young children attending day-care facilities. It is also common in travellers returning from holidays in the Himalayas and the former Soviet Union, where the water supplies of several cities appear to be chronically contaminated.

Balantidium coli is a protozoan often seen as a commensal of the human intestine. It is also frequently found in the intestines of pigs. It is being increasingly recognised as a rare cause of a dysenteric illness in humans who drink water contaminated with pig faeces. Likewise, *Cryptosporidium parvum* is a protozoan that is frequently isolated from cattle. Several cryptosporidia are associated with sheep and yet more can be isolated from birds. Like *Balantidium coli,* they are now being recognised as a significant cause of human diarrhoeal disease. Some people may be asymptomatic carriers; in others it causes diarrhoea. Epidemics of cryptosporidiosis have been attributed to the failure of domestic water purification systems. Severe, long-term diarrhoea caused by cryptosporidia is one of the most common presenting symptoms in patients with AIDS, although it can cause severe disease in anyone who is immunocompromised. The onset of chronic, protracted cryptosporidial diarrhoea often marks the transition from AIDS-related complex (ARC) diseases to full-blown AIDS.

Natural warm water springs form the ideal habitat for the amoeba *Naegleria fowleri.* This protozoan causes a very rare form of meningitis. The amoeba is inhaled and burrows through to the brain from inside the nose. Once in the brain, rapidly fatal meningitis ensues. Bathing in the Roman Baths in the City of Bath is no longer permitted because the water source for the baths is contaminated with *Naegleria fowleri*. This provides an unusual example of a water-borne infection that has exclusively extraintestinal symptoms.

3.2 How is water examined to ensure that it is safe to drink?

Because of the risk to human health caused by water-borne infections, in developed countries microbiological monitoring of water supplies is regularly undertaken. There are many water-borne infections. In consequence it is impractical to test water supplies for the presence of all the possible causative agents. Indeed, these organisms are often present in very small numbers and tend to die out rapidly. It is, therefore, the practice to test for the presence of **faecal indicator organisms**. These represent the inhabitants of the mammalian gut and are found only in the bowel, or in areas contaminated with faeces.

There are a number of desirable qualities required of a good faecal indicator. These include the ability to survive in all water sources that act as potential drinking supplies and to survive longer than the hardiest enteric pathogen. A good faecal indicator must also always be found wherever enteric pathogens occur. Furthermore, it should not be able to reproduce in contaminated water, otherwise an overestimate of the degree of faecal contamination of the water source will result.

Typically, three bacteria are used as indicators of faecal pollution: *Escherichia coli,* enterococci and *Clostridium perfringens.* Water supplies may be tested for the presence of all three bacteria but it is common practice to look only for *Escherichia coli.* The coliform group of bacteria can be found in the guts of a variety of animals and birds. Therefore, it is possible to isolate these bacteria from waters distant from human faecal contamination. It is also necessary to distinguish faecal coliforms such as *Escherichia coli,* derived almost exclusively from intestines, from other members of the Enterobacteriaceae such as *Klebsiella pneumoniae.* This bacterium may also grow in soil and upon vegetation. Its presence in water need not be the result of faecal contamination.

In the case of *Escherichia coli,* water supplies are first tested for the presence of coliform bacteria, including all the lactose-fermenting Enterobacteriaceae. These may or may not be of faecal origin. This test is referred to as the **presumptive coliform count.** To ascertain the numbers of *Escherichia coli* present in the original water sample, bacteria from the presumptive coliform count are subcultured in media in which only the faecal coliforms normally grow. This permits the **differential coliform count** to be determined. This provides an estimate of their numbers in the original water sample. Finally, to distinguish *Escherichia coli* from other faecal coliforms, the '**IMViC**' tests are used on colonies sub-cultured onto a solid medium from the differential coliform test cultures. *Escherichia coli* is typically positive for the production of indole and in the Methyl red test but it is negative for the Voges–Proskauer reaction and the utilisation of citrate.

Enterococci are more able to survive in brackish waters than are the faecal coliforms and they provide a more reliable indicator of recent faecal contamination in such sites. Enterococci are sought where there is difficulty in interpreting the results obtained from coliform counts. Failure to isolate enterococci does not, however, exclude the possibility that the water sample tested is contaminated with faeces.

Sporing anaerobes are highly resistant and their presence without other faecal indicators points either to distant faecal pollution, or the sporadic contamination of water occurring some time previously. As with faecal coliforms, a presumptive clostridial count is performed initially and subsequently, the presence of *Clostridium perfringens* is confirmed by the production of a **stormy clot** in litmus milk cultures.

There are two commonly employed methods of testing waters. When small numbers of organisms are expected, the membrane filtration technique is employed. This method gives good reproducibility and allows very large volumes of water to be examined easily, which greatly increases the sensitivity of the count. It cannot, however, be used to count bacteria in waters of high turbidity, nor in waters that have a large population of aquatic bacteria since these may overgrow faecal indicators. Furthermore, heavy metal ions and phenolic compounds present in the water may concentrate upon the filter material, inhibiting the bacterial growth and lowering the observed counts. Consequently, in some situations the more expensive and technically more demanding 'most probable number' (MPN) counting method is used (Fig. 3.1), exploiting a variety of selective media.

Using the most probable number counting technique, increasingly small volumes of water are used to inoculate the chosen broth medium. As the sample volume decreases, so too does the probability that the sample will contain a microbe capable of growing in the broth into which it is inoculated. The probability of such events can be calculated statistically using the Poisson distribution. This statistical distribution is used to predict the chance that a rare event will occur. By observing the number of cultures that yield positive growth after an appropriate incubation, reference to probability tables based on the Poison distribution will give the most probable number of microbes in the original sample.

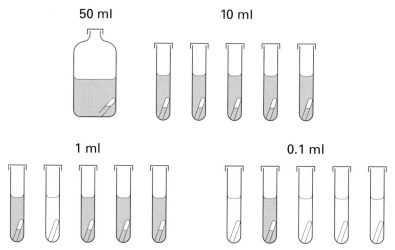

Fig. 3.1. Determination of the most probable number (MPN) of organisms in a water sample. One 50 ml sample, five 10 ml samples, five 1 ml samples and five 0.1 ml samples are inoculated into an appropriate broth. If coliforms are being detected, a Durham tube is included to trap any gas formed. After incubation, the numbers of samples yielding a positive reaction (in the case of coliforms, turbidity, acid and gas production) are counted and the most probable number of bacteria in the original sample is determined by referring to published MPN tables. In this case, the 50 ml sample is positive as are all the 10 ml samples, four of the 1 ml samples and one of the 0.1 ml samples. This equates to a most probable number of 170 coliforms per 100 ml in the original sample.

Although the bacteriological examination of water is regularly undertaken, the virological testing of drinking water is less common. Many viruses that cause water-borne diseases cannot be cultured readily in the laboratory. They may also persist in water for longer than bacterial faecal indicator organisms. The inoculation of test water samples into tissue cultures and the subsequent examination for plaque development or a specific cytopathic effect has been employed, but these tests are confined to specialised laboratories. To circumvent these problems, it has been proposed that bacteriophage may act as virus indicators, highlighting water that may potentially be infected with viruses. The practice of examining drinking water for viruses is not widespread. Perhaps this is an area where the application of molecular biological techniques such as PCR could prove to be of value. Care must be taken when interpreting the results of such tests, however. The presence of virus nucleic acid in a sample need not necessarily correlate with the presence of infectious virus particles. The European Community have a set of standards relating to the concentration of virus particles in waters used for recreational purposes.

Certain British waters currently fall sadly short of these ideals! Many Mediterranean beaches are also suspect in this respect.

3.3 How is water purified to ensure that it is safe to drink?

Rain water as it falls has, until recently, been assumed to be fairly pure. Industrial pollution has changed this assumption, most notably regarding the formation of acid rain. This is caused when industrial effluent gasses dissolve in atmospheric water droplets. From a microbiological point of view, however, rain water is relatively free from contamination. In contrast, the water found in soils, collected as surface waters, is always biologically active and may contain large numbers of microbes, as well as a high concentration of organic matter. Additionally, surface waters often possess a high concentration of mineral salts. Surface waters provide nutritionally rich environments that may support extensive and complex microbial communities. Water from soils percolates through the underlying porous rocks to collect over impervious rocks as ground water collections called **aquifers.** During its passage through the rocks, the mineral content of water increases but the microbial populations are significantly reduced to a point where ground water may not be biologically active. Surface waters act as a biological buffer between rain water and the water present in aquifers.

For centuries, natural springs and wells have provided a relatively safe supply of reasonably clean drinking water for human use. The often high mineral content may, however, give such waters an unpleasant taste and in some cases an unpalatable smell as well. Such waters were often exploited in one-time fashionable spa towns on the basis that if it was unpleasant to take then it must be of benefit. There is an unpleasant taste and smell associated with natural spa waters. This may be coupled with the small risk of contracting a water-borne illness from natural sources. With hindsight, these observations would provide a ready excuse for drinking ale rather than water, which at one time was popularly known as Adam's Ale. Beers have a pleasant flavour and the alcoholic component makes them microbiologically safer to drink than untreated waters.

The best water supplies in the past were either wells or fast flowing streams. In these water sources, the microbiological load of the water is minimised. If water is withdrawn from wells too rapidly, however, this will disturb the natural percolation process. An unacceptable microbial contamination of the aquifer may result. Today, human populations congregate in large numbers in conurbations. These do not always have adequate natural water supplies. Drinking water must be collected in large reservoirs and piped to supply our water demands, often over considerable distances.

The widespread use of fertilisers causes considerable problems for the water supply industry. These leech into rivers and reservoirs, where they cause algal blooms. The blooms not only result in the production of unpleasant tastes and odours in water but also provide organic matter that may support the excessive growth of water-borne pathogens. It is also now recognised that certain algae can produce toxins that produce diarrhoea in animals and humans. Some algal blooms may even prove fatal to cattle and domestic pets when they drink from contaminated waters. Pollution problems may also arise from industrial plants and from the discharge of toxic chemicals. Plants such as nuclear power stations discharge large quantities of hot water into rivers. This may profoundly affect the local microflora and consequently alter the whole ecology of the affected river.

Organic matter settles out of surface water to form a sediment within reservoirs. Particles flocculate in the water and sink, often taking microbes with them. This reduces the microbial population in the upper portion of the reservoir both by the physical removal of organisms and through nutrient depletion. Acid water tends to inhibit microbial growth in reservoirs. If, however, large populations of waterfowl visit the reservoirs, these may continually add to the microbial load by defaecating in the water. This is one way in which campylobacter infections may spread. Similarly, soiling of water by domestic animals may pollute reservoir waters. Algal blooms, confined to the upper levels of the reservoir, may also cause considerable problems in the reservoir storage of drinking water. Drinking water may be withdrawn from layers below the bloom but above the organic sediment.

The treatment of drinking water depends upon its source and the quantity and types of pollutant present, both chemical and biological. Waters that are relatively clear may simply require filtration before chemical treatment. Those of a high organic content may require coagulation and flocculation before filtration to remove the particulate matter that is present in the water. Coagulants used in the water supply industry include aluminium sulphate, sodium aluminate and ferric sulphate. These salts alter the forces on charged particles, causing them to flocculate. Care must be observed in the use of chemical additives for water treatment. Addition of excessive amounts of aluminium salts to the water supplies caused severe health problems for people living in Camelford, Cornwall, whose supplies were affected as a result of an accident in 1988.

Flocs grow in size and then sink under the influence of gravity in settling tanks. Clear water is run off the settling tanks and onto filtration beds. Coarse sand removes more particulate matter. Fine sands, gravel and anthracite beds are also used in the filtration of drinking water. This does not rely simply on the mechanical removal of microorganisms. Rather, it depends upon the activity of a **'schmutzdeke'** to be found in the upper layers of sand. This comprises a slime layer containing a rich population of bacteria, algae and protozoa. The bacteria in this layer may produce antimicrobial substances and the protozoa include scavengers that actively feed upon other microbes, including potentially pathogenic organisms. The schmutzdeke significantly reduces the biological load of the filtered water. Eventually the schmutzdeke accumulates to clog the filter bed. When this happens, a new sand bed must be used. Its water is not considered fit for consumption until *its* schmutzdeke is established.

Activated charcoal beds are used as a final filtration for water supplies that carry a high concentration of organic chemicals. These are adsorbed onto the surface of the charcoal, thus removing them from the drinking water. In the UK, water is then disinfected by the addition of chlorine. After treatment, a residual concentration of 3–5 parts per million is allowed to remain. This helps to prevent post-treatment microbial contamination through cracked distribution pipes etc. In other countries, ozone is used as a disinfectant in place of chlorine. This has the disadvantage that it does not form a residue in the treated water that can persist to the point of delivery and consequently it cannot prevent secondary contamination of the water. In its favour, however, is the fact that it avoids the risk of the formation and accumulation of chlorinated organic compounds in drinking water.

One of the greatest triumphs of the advertiser's art in recent years has been the enormous success of bottled water. Playing on public concerns regarding the purity of tap water, huge numbers of people have been lured into buying 'natural' bottled waters by the gallon. The argument is that because it is natural it must be pure. There is a dreadful irony here. The delivery of domestic water is very tightly regulated to avoid the possibility of water-borne infections. The bottled water industry appears to be less well regulated. Although the acid generated by carbonating waters effectively sterilises the fizzy water product, bottled water that is still can contain high bacterial counts. Although there have been no reports of bottled spring water containing bacteria of faecal origin, staphylococci have been isolated from water samples. This implies poor hygiene in the bottling process since staphylococci are part of the commensal skin flora. Furthermore, bottled spring waters can contain high numbers of pseudomonads. Although these pose no threat to healthy people, the immunocompromised are vulnerable to systemic, life-threatening infections caused by these bacteria. These are often resistant to a wide range of antibiotics and the infections they cause can be extremely difficult to treat. People who are immunocompromised are accordingly advised not to drink bottled water because of the risk to their health that this 'natural' product carries.

3.4   How is sewage treated to make it safe?

Human populations grow larger and larger. Communities grow in single locations. Population growth and concentration bring an increasingly difficult problem: how to dispose of human sewage? This is further complicated by the problems of disposal of industrial effluents. These frequently require chemical treatment before they can be discharged into the environment. This is required either to remove toxic chemicals from the water or to adjust the balance of nutrients available to assist the microbial degradation of organic wastes.

The increasing human population also has a rising demand for clean water. As a result, it becomes essential for the waste water from sewage to be purified and then returned to the natural water cycle. Consequently, the old methods of sewage disposal, either dumping it onto arable land or discharging raw sewage into rivers or out at sea, are becoming unacceptable. This is unacceptable not only because of the smells generated or the volume of the wastes involved but principally because of the risks to health.

The quality of sewage may be estimated by measuring the **biological oxygen demand** or **BOD** of the waste. This is defined as the quantity of dissolved oxygen used by a defined volume of waste, often 1 litre, after being held in the dark for a specified time and at a specified temperature. Today the most common temperature used is 20°C and the time of the test is most often five days. Samples are held in the dark to prevent the evolution of oxygen by photosynthetic organisms. The BOD acts as a rough guide to the active organic content of the waste. If the BOD of the effluent exceeds the oxygen concentration of the water into which the effluent is discharged, then the oxygen in the water will be rapidly used up, upsetting the natural ecological balance. Fish will die as a result and anaerobic bacteria will proliferate, generating noxious odours. Sewage must be treated to reduce its BOD and to render its solid wastes harmless. A variety of approaches to this problem have evolved.

In isolated rural dwellings, the simplest solution to sewage disposal is the **septic tank**. Sewage is collected in a large settling tank. Solid waste falls to the bottom of the vessel and forms a sediment where anaerobic digestion reduces the organic content of the effluent. Clarified fluid effluent may then be discharged: sometimes after further treatment.

An alternative method of sewage disposal found in small rural communities is the use of **oxidation ponds**. Settled sewage is held in ponds for upwards of 30 days. As the name implies, organic matter undergoes microbial oxidation in oxidation ponds. This converts waste compounds to inorganic compounds including water, carbon dioxide and ammonia. Algal growths are very important to the efficient operation of oxidation ponds as these generate oxygen as a product of photosynthesis. This is necessary for the breakdown of organic wastes. After processing sewage in oxidation tanks, very little solid residue remains.

Where larger populations gather, alternative sewage processing systems have developed to meet the increased demand for sewage disposal. Initial processing involves the screening of wastes to remove large solid matter. The resulting effluent is allowed to flow slowly over large settlement beds. Grit and stones present in the effluent settle out and flocculants are added to increase the precipitation and sedimentation of organic matter. The same flocculants are used in sewage disposal and in water purification. In the UK, settlement is allowed to continue for 15 hours. The liquid effluent, together with its suspended flocs, may then be treated either on **biological filters** or in **activated sludge plants**. The sediment from settlement beds requires anaerobic digestion before disposal to render it safe. Settlement can reduce the BOD of waste material by about 40%.

Biological filters comprise large circular tanks, about 2 metres in depth and filled with an inert material such as gravel. Their purpose is to provide a vast surface area to support the growth of the biological filter. Four arms rotate over the tank, distributing liquid effluent gently and evenly across the surface of the filter. A single plant can process 50 million gallons of sewage daily, but the flow through the system has to be carefully regulated. If too much material flows through the tank, the filter cannot function properly and the effect will be to discharge untreated sewage into the environment. Consequently, biological filter plants must be designed to accommodate extra capacity above the normal demands.

The predominant bacterium in the upper layer of the biological filter is the pseudomonad *Zooglea ramigera*. Its cells have many finger-like projections and they produce large quantities of slime. This microbe has the capacity to metabolise a wide variety of organic compounds. Bacteria of the genera *Nitrosomonas* and *Nitrobacter* oxidise ammonia, ultimately to produce nitrates. These are to be found in the lower layers of the filter beds. A variety of saprophytic **moulds** are to be found in filter beds feeding on decaying organic matter. Predatory motile protozoa move over the upper reaches of the tank, actively scavenging for bacteria and other microbes. Sedentary, stalked protozoa may be found devouring microbes lower in the tank. Many invertebrates graze upon the microflora of filter beds and these help to reduce the risk to the plant of clogging. The cleaned effluent runs from channels in the bottom of the tank. The BOD of sewage waste treated in biological filter beds is reduced by 85–90%: the reduction in solid wastes is slightly less. More than 97% of bacteria and viruses present in the original sewage are removed as a result of this treatment. People who keep aquaria will be familiar with biological filters, albeit on a much smaller scale. After all, fish living in aquaria are living in their own toilets.

Activated sludge tanks are also large-scale sewage treatment plants. Liquid effluent from settling beds is pumped into a large tank and the contents are aerated, either by vigorous stirring or by pumping compressed air through the container. During aeration, *Zooglea ramigera* encourages the formation of flocs and this helps in the sedimentation of solid waste following activated sludge treatment. Nitrifying bacteria have to compete with **heterotrophs**, hence the need for vigorous aeration of the activated sludge vessel. These bacteria fulfil the same function as in biological filter beds. Protozoa also feed on bacteria in activated sludge plants. This reduces the BOD by up to 90% and the solid matter is removed to a similar extent.

Activated sludge treatment is as efficient at removing bacteria and viruses as is the biological filter bed treatment. The liquid effluent remaining after activated sludge treatment is safe to discharge. Some of the solid matter that settles from activated sludge plants can be used as an inoculum for further activated sludge processes. Problems can arise when filamentous bacteria such as those of the genus *Beggiatoa,* the genus *Sphaerotilus* and the genus *Thiothrix* accumulate in activated sludge tanks. They cause the formation of loose flocs. These do not readily settle out to form a sediment when the activated sludge treatment is complete.

In both the biological filter beds and in activated sludge plants, Gram-negative bacteria are destroyed by *Bdellovibrio bacteriovorus*. This bacterium can penetrate the outer membrane of Gram-negative bacteria and lives in the periplasm as a parasite. This ultimately has fatal consequences for its hosts.

The sediments from biological filter beds and from activated sludge plants have been used directly as fertilisers. In contrast, the sludge generated by primary settlement tanks may contain pathogens in abundance. It also has a high concentration of nutrients. To render this material safe, it is placed into a digester. This is a large-scale septic tank where the contents are stirred and heated to 30–35°C. The sediments from biological filter beds and activated sludge plants may also be further processed in digesters to render them harmless.

In digesters a variety of clostridia are responsible for the initial breakdown of complex organic compounds that act as nutrients for these obligate anaerobes. This produces simpler compounds that are further broken down by a range of methanobacteria including *Methanobacterium formicum, Methanosarcina barkerii* and *Methanobacillus omelianskii.* These bacteria are responsible for the generation of considerable quantities of methane and this can be used as a fuel supply, furnishing all the energy demands of the sewage plant. Anaerobic digestion of sewage waste is also responsible for the generation of many unpleasant smells. After digestion has continued for up to one month, the liquid is filtered off and the solid waste is either burned or, after pasteurisation, is applied to the land as fertiliser.

Water from sewage treatment plants may be used for industrial processes or may be further treated. After aeration it may be discharged into rivers, lakes or the sea. It may also be chlorinated to disinfect it for subsequent use as a source of drinking water. By the time water has reached the East End of London, it may well have passed through several sewage plants.

4

Microbial products

Glancing through the contents of this book you could easily be forgiven for thinking that microbes were mostly malign, the cause of destruction, disease and death. Indeed, because of the vital importance of the control of infectious diseases, an enormous effort has been directed towards the medical applications of microbiology. Not even Louis Pasteur was lured into a study of microbiology with the intention of ridding the world of disease, as he later aspired to do. Rather, he was appointed to solve an economic and industrial problem threatening to ruin France.

Pasteur started his professional life as a chemist. As a young man, he had found that when polarised light is shone through solutions of organic compounds derived from living organisms the beam is either diverted to the left (laevo-rotatory) or to the right (dextro-rotatory). When organic compounds are synthesised *in vitro*, no such rotation is seen. He explained this phenomenon by showing that the L-molecules and the D-molecules were optical **isomers**. In nature typically only one isomeric form is produced or used by organisms. There is a notable exception to this. The bacterial cell wall contains peptidoglycan, a polymer incorporating sugars and both D- and L-amino acids. When organic chemicals are synthesised in test tubes, however, there is an equal chance that either isomer will be made. Because of the reputation gained from this work, in 1854 Pasteur was invited to the University of Lille to study the problem of beer spoilage. This led on to a study of wine production and spoilage. Pasteur demonstrated that wine production was not a chemical process. Rather it is a fermentation caused by **yeasts**. He also showed that the souring of beer and wine that was threatening much of the French drinks industry was caused by bacteria contaminating the wine and converting alcohol to acetic acid. This work led on to similar studies concerning the souring of milk. He found that this process could be delayed simply by heating milk for a short time to kill the bacteria responsible for causing souring. **Pasteurisation** had been born.

A decade later, Pasteur's attentions turned to another industry that was of crucial importance to France: the silk industry. The production of silk had been dramatically curtailed because silk farms had been struck by a disease of silkworms called pébrine. Pasteur showed that this was not only a contagious disease but it also had a heritable component and could be controlled by breeding from disease-free resistant lines of silkworms. These examples illustrate that microbiologists ignore the industrial applications of their subject at their peril. They also illustrate that, albeit unsuspected, microbes could underlie or undermine essential industrial processes. Since this recognition, microbes have increasingly been harnessed to promote human industrial endeavour. The explosion of biotechnology since the development of molecular biology has reinforced the importance of microbiology to industry.

4.1 How did microbes contribute to the First World War effort?

Perhaps one of the most profound of microbial industrial contributions to world history is the story of the production of acetone. This is one of the compounds that is essential for the manufacture of the explosive cordite. The traditional way of making acetone was to distil it from woods like beech, birch and maple. About 1% of the volume of wood yields acetone and the process was voracious in its appetite for timber. When the First World War began, the demand for acetone rose steeply. The Allied industry could not meet the demand, so an alternative source was urgently sought. Chaim Weizmann at the time was Reader in Chemistry at the University of Manchester. Although born in Russia, he came to England to escape religious persecution and became a British citizen in 1910. The British Government approached him to see if he could come up with a solution. He turned his attention to microbial metabolism. He discovered that a Gram-positive sporing bacillus *Clostridium acetobutylicum* could make large quantities of acetone using glucose as a starting material. This was cheap and much easier to obtain than the vast number of trees that would otherwise have been needed. Glucose is converted into pyruvate during glycolysis and pyruvate is then metabolised to produce acetone. As a bonus, this bacterium also produced butanol, another organic solvent. In return for developing a process that made an enormous contribution to the war effort, Weizmann was offered political honours. These he declined. Instead he was instrumental in the Balfour declaration, published in 1917. This pledged the British Government to support 'the establishment in Palestine of a National home for the Jewish People'. This paved the way for the foundation in 1948 of the independent state of Israel: its first president, Chaim Weizmann.

It is not just the Allies that used microbes to make the chemicals necessary for the production of explosives. The German Army used nitroglycerine as its favoured explosive and its manufacture relies upon a plentiful supply of glycerol. Before the war the Germans imported this chemical but a naval blockade of German harbours prevented its importation. Glycerol is a minor by-product made during alcohol fermentation by *Saccharomyces cerevisiae*. Ethanol is made under normal fermentation conditions by the action of alcohol dehydrogenase on the acetaldehyde. The purpose of this reaction is to regenerate nicotinamide adenine dinucleotide (NAD^+) but this reaction could be prevented by the addition of sodium sulphite to the fermentation reaction. The German chemist Carl Neuberg discovered that, as the cells still required NAD^+, in the presence of sodium sulphite *Saccharomyces cerevisiae* simply increased its production of glycerol. Later, an alternative biological source of glycerol was discovered. The **halophilic** alga *Dunaliella salina* produces large quantities of glycerol to act as an osmotic stabiliser. This helps the cells to survive the huge osmotic pressures generated in the high-salt environment in which these organisms live. Currently, organic solvents are made not by biological fermentations but rather as by-products of the oil industry. Oil is a fossil fuel and as a consequence it is not a renewable resource. The time may soon be approaching when the biological fermentation reactions will again become economically viable for the production of organic solvents.

4.2 What role do microbes play in the oil industry and in mining?

The oil and petroleum industry is currently exploiting a finite resource to provide the raw material for numerous other industries. Oil products are used in the pharmaceutical and food industries, in the production of fabrics and fertilisers, to make plastics, paints and building materials and in the generation of electricity. Indeed, modern developed societies are highly dependent upon the oil industry and its offshoots. Oil is recovered from shale: a sedimentary rock in which chemical and physical reactions have acted upon organic matter, mainly from decomposing marine organisms, to produce a mixture of hydrocarbons. Natural gas is generated by the decomposition of organic matter in oil shales to produce volatile, short-chain hydrocarbons. Methane, the gas of North Sea gas, has a single carbon atom: the shortest chain possible. Natural gas rises above the oil in shales to be trapped by overlying rock strata. Oil shales also carry liquid oils and waxes: solid products formed from long-chained hydrocarbons.

The presence of gas under pressure allows the liquid oil to be extracted easily, at least in the early life of an oil well. When a bore hole is drilled into an oil reserve, the pressure of the expanding gas pushes the light liquid oil out of the well. No more than about 25% of the oil in a given deposit may, however, be recovered in this way. Pumping water into the well can extract much more oil. The oil will float on the water and can thus be forced from the well. This increases the yield from the well, allowing extraction of up to 60% of the deposit. The waxes in shales cause serious problems since they can clog the pipes and are difficult to extract. The waxes found in oil shales are paraffins. Pumping steam into wells rather than water can melt some waxes, relieving the problem to some extent. There are experiments under way to enhance still further the recovery of oil from deposits by using microbes that produce surfactants. Xanthan gums made from the pseudomonad bacterium *Xanthomonas campestris*, for example, are widely used to help wash oil from shale. These are pumped into oil deposits to help the oil extraction process. Other microbes are also used to digest the long-chain solid paraffins. These microbes not only assist in recovering lighter oil fractions but they also prevent solid waxes from building up within the bore pipes, thus blocking the bore holes.

The xanthan gums as used in the oil industry have amazing properties. These make them very widely exploited. They are also used as lubricants and they can form smooth emulsions that are not disturbed over a wide temperature range. This means that they are used to make paints and ceramic glazes. They are non-toxic and are widely use in the pharmaceutical industry. They are also incorporated into processed foods such as ice cream, fruit drinks and salad cream. They are a truly rich resource.

The microbes used in the enhanced recovery of oil are similar to those that can help to degrade oil spills. The oil industry relies upon extracting oil from isolated locations and transporting crude oil, often over considerable distances. Oil tankers regularly move vast quantities of crude oil over the world's oceans and almost inevitably accidents happen. On a minor scale, ships often discharge crude oil at sea when washing out their tanks. Occasionally, however, oil tankers are seriously damaged at sea and major oil spills can cause environmental disasters. Work on the Exxon Valdez disaster in Alaska in 1989 illustrated the importance of microbes in recovering the natural environment following major crude oil spills. The lessons from this ecological disaster have been subsequently applied to the **bioremediation** of other major oil disasters. One such disaster occurred on 21 February 1996 when the Sea Empress spilled 70 000 tonnes of oil into the sea around southwest Wales. This was almost double the size of the spill of oil from the Exxon Valdez.

Inorganic nutrients containing nitrogen salts and phosphates are applied to spilled material and this encourages the growth of microbes capable of digesting oil. Crude oil is a complex mixture of hydrocarbons, and different microbes are capable of attacking different components of the oil. As the crude oil is metabolised, various microbial products are generated and these may act as substrates encouraging colonisation by additional microbes. Consequently, the microbes that can attack crude oil spills are a complex and constantly evolving community. Important microbes in the bioremediation process include pseudomonads, mycobacteria and the aerobic coryneform bacteria. The breakdown of crude oil is an aerobic process. If oil could be metabolised anaerobically then the deposits in oil shales would have disappeared long ago. This, clearly, has not happened. Whereas the microbial breakdown of oil in a bioremediation process can bring enormous environmental advantages, the microbial digestion of oil and oil products can cause problems when storing petroleum. Lead was added to petrol not simply to help it burn better in older engines; it also acted as an antimicrobial agent.

Oil is also transported through pipelines. The delivery of organic matter in metal pipes may be problematic, with microbes playing the villains. The microbial attack of metal pipes also causes severe corrosion problems in oil fields where water is used in the secondary extraction of crude oil from wells. Bacteria of the genus *Desulfovibrio* can, under anaerobic conditions, reduce the small quantities of sulphate present in oil deposits and the sulphide produced reacts with the iron in the pipes causing corrosion of the metal. Methanogenic bacteria also contribute to this process. These are very ancient bacteria that can use elemental iron as a source of electrons for their metabolic requirements. This microbial corrosion causes significant problems for any iron pipes buried underground in potentially anaerobic conditions. Domestic water and gas pipes were once made from cast iron but the problems caused by microbial iron erosion means that they are now being replaced by plastic piping. It has been estimated that up to 40% of domestic water is lost through leaking pipes before it reaches the consumer.

The world oil reserves are finite. The first oil crisis was created in the West in the early 1970s when the Gulf States began to ration sales of oil. This demonstrated its enormous importance as an economic resource. We will have to face a time when the supplies of oil are exhausted. Coal may provide an alternative source of organic fossil fuel, but our coal supplies are also finite. The search is on for renewable energy sources but this will only provide part of the answer. Oil is not just used as a source of energy. It is the starting material for many production processes and an alternative supply of hydrocarbons must be maintained to avoid a necessary revolutionary change in human lifestyle, at least in developed countries.

Stromatolites are mats of filamentous microbes, and there is evidence that these were common in areas that later developed into the oil shales that we exploit today. Consequently, microbes probably played a significant contribution to the original formation of our oil deposits. It is possible that, properly harnessed, microbes may be able to meet our demand for hydrocarbons today. When the demand for oil rose because of the Industrial Revolution and its aftermath, great efforts were expended in oil exploration. About 100 years ago, a bituminous substance was found in the Coorong region of South Australia. This was named 'coorongite' and was thought to have leaked from underground oil reserves. Coorongite was rich in hydrocarbons, but despite extensive exploration no underground oil reserves were found in the area. It was only later that coorongite was found to have formed from the growth of a colonial alga, *Botryococcus braunii*. Up to 30% of the dry weight of this microbe comprises petroleum-like long-chain hydrocarbons when it is grown under suitable conditions.

One of the major uses of petroleum other than as a fuel is to provide the starting material for the plastics industry. Because of the ease with which they can be manipulated and their mechanical properties, plastics have found a huge array of applications. However, many plastic products are considered to be 'disposable' and it has been estimated that 40% of plastic products made annually are disposable and will end up in landfill sites. Plastics are **xenobiotic** compounds. These are artificial compounds that do not naturally occur on the planet. As yet, microbes have not evolved to degrade plastics. These probably represent the ultimate challenge for microbial metabolism. Interestingly, other xenobiotic compounds, including pesticides that are designed to be toxic, can be metabolised by microbes. This may be a slow process and to effect the complete biodegradation of certain recalcitrant chemicals the cooperation of more than one microbe may be required. Often pseudomonads and environmental mycobacteria are involved in the metabolism and biodegradation of xenobiotic compounds. There is evidence that pseudomonads, in particular, can evolve to meet the challenges presented by xenobiotic chemicals. The genes encoding enzymes capable of metabolising unusual chemicals may be found on plasmids. One of the best studied is the TOL plasmid of pseudomonads, which codes for the metabolism of toluene.

The volume of waste plastic products we currently generate is causing a serious long-term problem. Consequently, there is considerable interest in developing natural polymers that share the desirable properties of plastics while at the same time are biodegradable. Microbes often generate polymers as energy storage compounds. The metachromatic volutin granules of corynebacteria stain pink rather than blue when treated with methylene blue. They are made of β-hydroxybutyrate polymers. These were amongst the first biopolymers that act as storage compounds to be extensively studied. Other microbes form different energy storage polymers. Poly-β-hydroxyalkanoates are microbial energy storage polymers that have the qualities required for a bioplastic. These polymers comprise repeating units of β-hydroxybutyrate, based on a four-carbon repeating monomer, and β-hydroxyvalerate, with a five-carbon monomer. *Alcaligenes eutrophus* is currently the most important source of bioplastic based on poly-β-hydroxyalkanoate. It can yield up to 80% dry weight polymer material.

Besides finding an alternative source of hydrocarbons for plastics and other organic chemicals, alternative sources of fuel will have to be found when the oil supplies dry up. Although clean, renewable sources of energy such as wind and wave power are now beginning to be exploited, these can meet with considerable opposition. Plans to site wind farms on the Yorkshire Dales overlooking Howarth have caused outrage to lovers of *Wuthering Heights*, who claim that the bleak moors that inspired this work will be ruined. Microbes could yet again provide an answer to the energy problem.

There have been numerous reports of veterinary surgeons causing explosions when releasing gas from the stomachs of cows. This results from accumulation of methane in the rumen. The methane produced by the microbial digestion of sewage is sufficient to run generators that provide for the electricity needs of the entire sewage works. In areas of the world where sugar is cheaply and plentifully grown, alcohol can be generated by fermentation and distillation processes and this alcohol can be used to run suitably adapted internal combustion engines. At the start of the information technology revolution, people believed in the 'paperless office'. Most people, however, still prefer to read paper copies of reports etc. With the widespread use of word processors, the demand for paper has, ironically, increased. This generates a mountain of waste paper: a cellulose-based product. Considerable effort is currently being directed at ways of converting the sugars locked in cellulose polymers into a form that can be used for economic fuel production. In the process, this will help to reduce the increasing waste paper problem.

Just as microbes cause problems with subterranean pipes, so they can be exploited to extract minerals from low-yielding ores. The best example is in the mining of copper, but other metals, including cobalt, manganese, nickel, uranium and zinc, can be retrieved using similar processes. The chemotrophic *Thiobacillus ferrooxidans* is used in the extraction of copper. Bacteria are mixed in solution and allowed to percolate through low-grade copper ores. They can metabolise sulphides in the mineral to produce an acidic sulphate. This, in turn, reacts with the copper and it dissolves as copper sulphate. The copper sulphate solution that elutes from the ore can then be harvested to yield elemental copper. As much as 10% of the copper produced in the USA is extracted with the aid of microbial metabolism.

Another application of microbes in the mineral industry is using immobilised microbes in the recovery of waste. There are soil fungi that can accumulate very high levels of cadmium metal. These have been grown as immobilised mats and used to extract cadmium from the exhaust gasses emitted from cadmium smelters. Cadmium is a highly toxic metal and thus the fungi are helping to reduce pollution from the smelter. They can accumulate upto 30% dry weight cadmium metal. Once a filter is exhausted, it is replaced and the fungi from the old filter are harvested. These are then placed in the smelter for the cadmium to be re-extracted. The fungal filters have a higher cadmium content than the cadmium ore normally used to feed the smelter. In a similar fashion, immobilised cells of the alga *Chlorella vulgaris* can be used to extract gold from streams. Other precious metals may be extracted by similar means.

4.3   How are microbial enzymes exploited?

Immobilised microbes may be used to extract minerals from dilute solution by acting as biological filters. This does, however, require that the microbial cells must be kept alive. Processes are now being developed to exploit not whole cells but immobilised microbial enzymes. These are being applied to a number of different industrial processes. Enzymes may be immobilised on beads or in resins, but often they are attached to the insides of tubes down which the liquid substrate can pass. Immobilised enzymes are used, for example, in the modification of naturally occurring antibiotics to produce a range of semi-synthetic drugs with different pharmacological and antimicrobial properties. They are now also widely used in the food technology industry in the production of amino acids and their derivatives, to reduce the lactose level in milk and to convert glucose to fructose in the preparation of high-fructose products.

Biosensors represent an exciting development in immobilised enzyme technology. Early attempts to develop biosensors used whole cells, but as the technology has improved, extracted enzymes are now being used. Whole bacteria have been used to detect metabolic poisons. *Photobacterium phosphoreum* uses ATP to emit light. These bacteria grow as saprophytes on rotting fish, causing the flesh to glow in the dark. Light emission is a very energy-demanding process and the quantity of light produced is directly proportional to the amount of ATP available to the bacterium. The light emitted can easily be measured by a photodetector. Exposing these bacteria to metabolic poisons will prevent ATP production. In turn, this will reduce the amount of light that the bacteria can emit.

In newer biosensors, microbial enzymes are attached to electrodes. As the enzyme reacts with a substrate, physicochemical changes occur that can be converted to electronic signals. In turn, these may be amplified and detected. By taking an array of sensors and using computers to analyse the output, very sensitive detection devices can be developed. Biosensors have been developed, for example, that can distinguish different **serovars** of salmonella when growing in pure culture. When this technology is developed it could form the basis of novel, non-invasive diagnostic techniques, 'smelling out' infections. Besides their medical uses, biosensors will have enormous applications in environmental management and in pollution control.

Enzymes are being exploited in ever more ingenious ways. Microbiologists are also becoming more aware of the importance of microbes that live in extreme environments. These require special properties to survive and they have evolved enzymes that can function under quite different conditions from those normally considered compatible with a familiar, comfortable life. Scientists are now beginning to exploit the properties of enzymes from extremophiles: the microbes that live in extreme conditions. Biological washing powders are one of the best examples of the exploitation of enzymes isolated from extremophiles. The detergent used in washing powders raises the pH of the wash to pH 9–10: considerably higher than many microbes can tolerate. Furthermore, washing is often carried out at an elevated water temperature. Although woollen and other delicate materials are washed at a temperature of 40°C, heavily soiled whites may be washed at 95°C. Many brands of washing powder now available come as a 'biological' product. These powders include a variety of proteases, amylases, lipases and other enzymes isolated from extremophiles capable of growth at elevated temperature and pH. These enzymes help to digest 'difficult' stains and are often obtained from bacteria of the genus *Bacillus*.

4.4 How do microbes help in the diagnosis of disease and related applications?

Who would credit that bacteria living in hot water streams could possibly be of more than academic interest? At first sight they would appear to be a mere curiosity. Indeed, they are very curious but they have recently been exploited in a way that has revolutionised molecular biology. PCR involves the amplification of short sequences of DNA. It has enormous applications and is a highly adaptable technique. It can be used to amplify very specific regions of DNA. Alternatively, by manipulating the reaction conditions, it can be used in a random amplification. In theory, a single molecule of DNA can act as a template and even in practice a very small number of molecules can successfully be amplified. This makes PCR technology very sensitive.

In essence, double-stranded DNA is heated to separate the two strands. The reaction mixture is then cooled and short oligonucleotide primers anneal to target sequences. For specific PCR reactions, primers are designed to be complementary to the target DNA, with one primer at either end of the target sequence. In non-specific PCR, the annealing conditions are adjusted, for example to allow pairing of non-complementary sequences. A DNA polymerase then synthesises new DNA extending from the oligonucleotide primer and using the single-stranded target DNA as a template. This process of strand separation, primer annealing and DNA replication is repeated, typically for 25 to 30 cycles (Fig. 4.1). This produces an amplified sequence of DNA that can be easily detected. To amplify RNA sequences, such as the genomes of many viruses, then the sample material is first treated with **reverse transcriptase**. This enzyme makes a DNA copy from an RNA template. The copy DNA can then be used for PCR amplification.

Having amplified a target DNA sequence, the product can be easily visualised by electrophoresis of the DNA on an agarose gel. This technique relies on the fact that DNA is a charged molecule that will migrate in an electrical field. The smaller a DNA molecule is, then the faster it will move through an agarose gel. DNA molecules can thus be separated according to size. This is a preliminary check to see if the correct product has been amplified. If the predicted sequence of an amplimer is known, a full nucleotide sequence determination can be carried out to ensure that the correct target has, indeed, been amplified. Alternatively, the amplimer can be digested with restriction

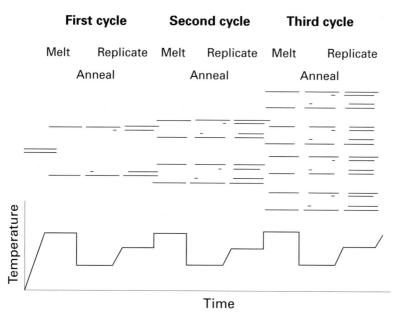

Fig. 4.1. The first three cycles of the polymerase chain reaction. The sample is heated so that the DNA template is melted. The reaction temperature is dropped to allow primers to bind, one to each strand, at either end of the target sequence. The reaction temperature is then increased to permit DNA replication mediated by the *Taq* polymerase. This reaction cycle is repeated for 20 or more cycles. By the end of the third cycle it can be seen that the majority of products are of a size defined by the limits of the primer-binding sites; hence when amplification is complete the product overwhelmingly contains DNA molecules of a particular size.

endonucleases. These are enzymes that cut DNA at specific and predictable nucleotide sequences. The resulting fragments can be separated electrophoretically and their sizes ascertained by reference to DNA size standards. This is a much cheaper alternative to nucleotide sequence analysis and is quite often used to confirm the identity of a PCR product.

When PCR was first introduced, a polymerase from *Escherichia coli* was used. At each strand separation step, the temperature required to melt the DNA denatured the enzyme. Consequently, before each replication step, fresh enzyme had to be added to the reaction. At every step, the reaction tube was opened and there was the possibility of contaminating the reaction. Either inhibitors would be introduced, giving false-negative reactions, or extraneous template DNA could enter the reaction leading to false-positive reactions. This is where thermophilic bacteria can supply a solution to the problem. The temperature required to melt DNA strands efficiently in PCR is typically around 95°C. The annealing temperature used in specific PCR depends upon the nucleotide sequences of the primer pair but it is typically about 60°C. The DNA polymerase of the thermophilic bacterium *Thermus aquaticus*, known as *Taq* polymerase, has a temperature optimum of 72°C. This lies neatly between the melting temperature and the annealing temperature. As the enzyme is purified from a thermophilic bacterium it is not degraded at 95°C. Consequently, a single application of enzyme at the start is all that is needed for a complete PCR amplification. Using an enzyme with a high optimum temperature also avoids the problems caused by non-specific annealing of primers to unrelated targets. This was a particular problem in the early days of PCR when the replication steps were carried out at 37°C.

PCR technology has an enormous array of applications. These range, for example, from forensic science to the taxonomy of microbes that cannot be artificially propagated. It is increasingly being used in clinical laboratories to diagnose infections that are difficult to discern using traditional technology. The diagnosis of tuberculosis is a good example of the application of PCR technology. It can take up to eight weeks to cultivate *Mycobacterium tuberculosis* in the laboratory. Using PCR technology, the diagnosis can be available within a single working day. There are many other conditions where PCR-based diagnostic tests are being developed. These include applications in the diagnosis of leprosy, Lyme disease, chlamydial infections and a range of virus infections where the virus cannot be cultured in the laboratory. PCR technology has found many uses and is also used in the diagnosis and epidemiology of genetic diseases. Perhaps one of its most spectacular successes, however, has been in the field of forensic archaeology, where the technique has been used conclusively to confirm the identity of skeletons of the Romanovs: the last Russian Royal Family, who were shot during the Russian Revolution.

Besides using revolutionary techniques such as PCR technology to diagnose infections, older methods are being refined as well. Serological tests have long been used in clinical laboratories. Either **antigens** have been used to detect **antibodies** raised by a patient in response to disease, or antibodies have been used to test for specific antigens in a specimen or culture. Antibodies can be made by purifying antigens and injecting these into the animal in which the antibody is to be raised. Although a single antigen is used in this method, the animal will have raised antibodies to other antigens it has met. Consequently, any serum from that animal will contain **polyclonal antibodies**. These days, **monoclonal antibodies** may be produced. These are pure preparations that will react exclusively with a single **epitope**. A mouse is immunised as in the production of polyclonal antibodies. When it has mounted an immune response, it is sacrificed and the spleen removed and its cells are treated with polyethylene glycol. This enables the spleen cells to fuse with other cells. They are then mixed and fused with **myeloma** cells. These are tumour cells that can divide indefinitely in tissue culture. The resulting **hybridoma** cells produce large quantities of antibody, each cell producing just one type of antibody. These cells are then grown and separated into clones: each clone producing a monoclonal antibody that will react with just one epitope.

Because of their high degree of specificity, monoclonal antibodies are increasingly used in the serological diagnosis of infection. They can also be used to study the epidemiology of infections. An array of monoclonal antibodies can be raised against various microbial epitope markers. The pattern of reaction of isolates with the collection of monoclonal antibodies will demonstrate the distribution of different pathogen strains during an epidemic of disease. The applications of monoclonal antibodies are not confined to infectious diseases. They can be used to identify tumour markers in the diagnosis of cancer and are used to classify leukaemias. Monoclonal antibodies are now used in the food industry, for example for the detection of mycotoxins such as aflatoxin and ergot in contaminated seeds and grain. Monoclonal antibodies may also be used to detect pesticides and antibiotics.

4.5 How do microbes contribute to the pharmaceutical industry?

Along with the exploitation of microbes and their products for diagnostic purposes, biotechnology is now increasingly being exploited to assist in the treatment of disease. Historically, the most important example is the use of vaccines to prevent infectious diseases. Vaccination derives its name from *vacca*, the Latin word for cow. Edward Jenner, a country doctor from Berkeley in Gloucestershire, had noticed that milkmaids who suffered from cow pox infections did not suffer from smallpox. He reasoned that the mild cow pox infection subsequently protected the milkmaids from the far more serious and potentially fatal smallpox. To test his theory, he took infectious material from a cow pox lesion and deliberately inoculated James Phipps, an eight-year-old boy. Six weeks later, Jenner inoculated him with material from a smallpox lesion but on this occasion James Phipps did not become unwell. Thus in 1796 vaccination was born. Jenner repeated his success and vaccination rapidly became popular as a method of preventing smallpox infection. In 1977 the last reported case of smallpox occurred and in 1979 the World Health Organization declared the disease eradicated. No other infectious disease has yet been eradicated by a vaccination programme, but by 1994 poliomyelitis had been eliminated from the American continent.

Several strategies have been adopted to provide immunological protection from disease. Vaccination against smallpox presents an unusual case. One virus, the cause of a mild infection, is used to provide protection against another virus capable of causing a life-threatening infection. The antigens on both viruses are sufficiently similar that antibodies raised against one may cross-react with the antigens of the second. The variola virus is used for small-pox vaccination. Its precise nature is still debated but it is thought probably to represent a hybrid between the cow pox virus and the smallpox virus. The Bacille–Calmette–Guérin, BCG, used to vaccinate against tuberculosis is another example of a non-pathogenic microbe affording protection against a pathogen, in this case *Mycobacterium tuberculosis*.

In some cases, killed pathogens are used to stimulate antibody formation. The first poliomyelitis vaccine, developed by Jonas Salk, was one such vaccine. Alternatively, pathogens may be treated so that they can no longer cause disease. This process is called **attenuation**. Repeated sub-culture in the laboratory will often weaken the pathogenic properties of an organism to the point where it no longer causes disease. Albert Sabin developed a live, attenuated polio vaccine. Because the poliomyelitis virus infects the gut, when it was first introduced this vaccine was delivered on a sugar lump, rather than using a needle. This was a very popular development.

Some microbial diseases are caused by the action of toxins. Early attempts to prevent development of toxin-associated disease involved the use of anti-toxin serum. Patients who have recovered from the disease will have antitoxin antibodies in their serum. This may be collected and administered to a patient in the early stages of the disease. Consequently, the antibodies in the serum provide a passive immunity, protecting the patient from further developing the disease. Passive immunity obtained from the use of antiserum is only short-lived, whereas active immunity will last much longer. It also carries the risk of transmission of blood-borne infections such as hepatitis and AIDS. Protection against such diseases may now be afforded by active immunisation with a **toxoid**. To make a toxoid, the relevant toxin is purified and then treated to render it harmless, while retaining its antigenic structure. Toxoids may be made by heating the toxin or by treating it with a chemical such as formalin.

A recent and exciting development in vaccination has been the use of **subunit vaccines**. Influenza virus undergoes continuous antigenic alterations. The influenza vaccination programme depends upon predicting the antigens likely to be circulating in the forthcoming months and preparing a cocktail of purified antigens to be delivered in the vaccine. Since purified antigens, free from infectious particles and nucleic acids, are used such vaccines are known as subunit vaccines. The hepatitis B vaccine is another notable subunit vaccine. In this case the surface protein of the hepatitis B virus has been cloned using recombinant DNA technology. This cloned protein is used to make the subunit vaccine. Another development has been the experimental use of cloned DNA as an immunogen for inducing protective responses. This area of DNA vaccination has exciting potential for the future design of vaccines to protect against virus infections.

Recombinant DNA technology is exploited to provide treatments as well as in the production of vaccines. Diabetics require a regular supply of insulin for injection to control their blood sugar levels. Traditionally, insulin has been purified from pigs and this has caused problems for various religious groups for whom pigs are unclean. Furthermore, the long-term use of non-human insulin can cause the diabetic to make antibodies to the foreign protein. Insulin was the first protein to have its amino acid structure determined. Biosynthetic human insulin was the first medically significant product of the biotechnology industry. The amino acid sequence of human insulin was used to derive a potential nucleotide sequence that could produce human insulin. Rather than searching for the human gene and cloning that, a predicted nucleotide sequence capable of encoding human insulin was derived from the published amino acid sequence. This nucleotide sequence was synthesised *de novo* using a series of controlled chemical reactions and the synthetic gene was cloned into the bacterial plasmid pBR322. The recombinant plasmid was then used to transform a laboratory strain of *Escherichia coli* and this was used for the source of biosynthetic human insulin. This overcomes the immunological and religious problems posed by pig insulin, but some diabetics find that when they use biosynthetic human insulin they are less aware when the levels are dropping than when pig insulin is used. This can cause problems for the management of the disease in these individuals.

There are other applications for recombinant DNA technology. Pituitary hormones are important in regulating growth and fertility. At one time, these precious products were made by collecting human pituitary glands at *post mortem* examination and extracting the hormones from the pooled material. It took pituitary glands from many people to make sufficient hormone for one person to be treated. The use of pooled organs carried the risk that when the hormone was purified so too was the agent that causes Creutzfeldt–Jakob disease. There have been cases of people given growth hormone or certain fertility treatments who have subsequently developed this fatal neurological disease because they received infective material together with their hormone therapy. Pituitary hormones such as human growth hormone and the gonadotrophic hormones are now made using recombinant DNA technology. This avoids the danger of Creutzfeldt–Jakob disease associated with the use of hormones prepared from human tissues.

Interferons are proteins produced by the body, particularly in response to virus infection. Clinical trials have demonstrated that administration of interferon can have a beneficial effect on the course of virus infections, including hepatitis and AIDS. Interferon therapy may also benefit patients with multiple sclerosis. It has been suggested that interferons may assist in the management of cancers but this is still controversial. Like hormones, interferons are naturally produced in tiny quantities. To overcome the supply problem, recombinant DNA technology is now used to meet the demand for these precious proteins.

The human body exploits two classes of hormone: peptides and steroids. Steroids are based on the same four-ring structure and the many minor modifications to the basic structure of steroids can profoundly influence the properties of the molecule. Steroid drugs have numerous applications in medicine. They can be used to combat organ rejection in transplant patients and to ameliorate the effects of immunological diseases such as asthma. They can be used for fertility treatments and as contraceptives. They also have a controversial role in building muscle tone. The synthesis of steroid hormones is complex. The fungus *Rhizopus nigricans* is used to cause a key specific hydroxylation necessary in the conversion of progesterone to hydrocortisone, one of the most widely used steroids. Exploiting natural microbial metabolism can often significantly reduce the number of synthetic steps necessary to modify steroids, making the production process more efficient.

Perhaps one of the most important uses of microbes in the pharmaceutical industry is in the production of antibiotics. After ulcer treatments and drugs used to control high blood pressure, antibiotics are among the most frequently prescribed drugs. Penicillin was the first antibiotic to be produced on an industrial scale. The first species of fungus used to make penicillin was *Penicillium notatum*; other species were later found to yield greater quantities. Today, strains of *Penicillium chrysogenum* are used in the manufacture of penicillins. Although Joseph Lister first recognised the antimicrobial therapeutic properties of penicillin, he could not obtain sufficient material to make its exploitation viable. Alexander Fleming never fully appreciated the therapeutic potential of this drug. When Howard Florey and his group first demonstrated the clinical application of penicillin, the UK was at war. Further developmental research was transferred to the USA. This was a happy coincidence since the Americans favoured the use of corn-steep liquor to grow fungi. This is rich in precursors for penicillin and the yield of drug was significantly improved when this liquor was used in the growth medium.

The first penicillin to be produced commercially was benzylpenicillin, penicillin G. This drug is very unstable in acid conditions and because of the stomach acid it can only be administered by injection. The first successful modification to the industrial production of penicillin was to incorporate phenylacetic acid into the growth medium. When fed with this precursor, strains of *Penicillium chrysogenum* make phenoxymethylpenicillin, penicillin V, rather than benzylpenicillin. This drug is much more stable under acid conditions and so can be taken orally without being significantly broken down by the acid in the stomach. By further manipulating the growth conditions, fungi will produce 6-amino-penicillanic acid. This is the active nucleus of the penicillin family and from it can be made a staggering variety of semi-synthetic penicillins with a range of antimicrobial and pharmaceutical properties.

A diversity of fungi and bacteria produce antibiotics. Of particular economic importance are fungi of the genus *Penicillium* and the genus *Aspergillus*, but most antibiotics are bacterial metabolites. Commercially, the most important producers are the streptomycetes, but the genus *Bacillus* is also a significant source of antibiotics. Other bacteria may become important in the future. One example is the Gram-negative genus *Erwinia*: source of the broad-spectrum antibiotic imipenem. This drug is used to treat life-threatening infections where other therapies have failed. Although most antibiotics are still made from microbes, simple structures such as chloramphenicol are more cheaply manufactured using an entirely synthetic process.

Vitamins are the most important non-prescription medications. Most are made economically using standard chemical syntheses but some are more easily produced biosynthetically. Yeast extracts such as Vegemite and Marmite are rich in vitamins of the B family. These tend to have complex structures. Microbes may be used to elaborate precursors used in the chemical synthesis of vitamins A and C, and the B family of vitamins are still manufactured using entirely microbial processes. Using sugarbeet molasses as a growth medium, *Pseudomonas denitrificans* is used to produce vitamin B_{12}. Alternative processes exploit members of the genus *Propionibacterium* to produce this vitamin. Although many bacteria and fungi produce riboflavin, vitamin B_2, the fungus *Ashbya gossypii* produces this vitamin in huge quantities and it is this organism that is commercially exploited to make vitamin B_2.

4.6   How do microbes contribute to food technology?

Citric acid, as well as being a chemical buffer widely used in the pharmaceutical industry, has many culinary uses. It is used in soft drinks, jams, sweets and a variety of other foodstuffs. For many years citric acid was extracted from lemons and the Italian lemon groves held a virtual monopoly on its production. These, however, could not meet the rising demand worldwide for citric acid and an alternative source was sought. Usually, citric acid is metabolised as part of the tricarboxylic acid cycle, the source of substrates used in aerobic respiration. Indeed, an alternative name for this metabolic pathway is the citric acid cycle. When growing on cheap substrates such as sugarbeet molasses, the fungus *Aspergillus niger* excretes huge amounts of this acid. Other substrates have been used including cane sugar and potato starch. Originally, the fungus was grown floating on its growth medium. This is the natural growth of an **obligate aerobic** mould. Higher yields of citric acid were obtained when submerged cultures were developed. This involved the design of so-called fermentation vessels in which the growth medium was agitated to distribute the fungal mycelia throughout the vessel. Because aspergilli are obligate aerobes it is necessary to ensure the fermentation vessels are adequately aerated throughout if citric acid is to be produced efficiently. The development of submerged aspergillus cultures paved the way for the later development of fermenter cultures used in the production of antibiotics.

In Chinese cookery, monosodium glutamate is a very popular flavour enhancer. People who are unaccustomed to eating this seasoning can suffer some discomfort, typically experiencing a hot flush. Nevertheless, it is widely used in cookery. Other amino acids are also used as food additives and the food industry has created a considerable demand for amino acids. This has been further enhanced by the search for low-calorie sweeteners as an alternative to sugars in 'diet' soft drinks and foods. One of the most popular artificial sweeteners is aspartame, produced from aspartic acid and phenylalanine. Most amino acids are most easily made by chemical synthesis but for some applications it is necessary to use the correct optical isomer. When chemically synthesised, equal quantities of the D- and L-isomer are made. To circumvent this problem, microbial fermentation or microbial enzymes can be used in the manufacturing process.

Corynebacterium glutamicum is used in the production of glutamic acid. Under normal circumstances, metabolites such as amino acids are not excreted. As the intracellular concentration of product accumulates, natural feedback mechanisms shut down the metabolic pathway responsible for synthesis. Biotin is a cofactor essential for lipid synthesis in bacteria. By growing *Corynebacterium glutamicum* on limiting amounts of biotin it is possible to make the bacterial membrane sufficiently leaky for the glutamic acid it is producing to escape from the cell. It cannot, therefore, accumulate sufficiently to cause feedback inhibition of the biosynthetic pathway. An alternative control is seen in the industrial biosynthesis of lysine. In this case, the bacterium responsible for production is *Brevibacterium flavum*. Mutants that are no longer susceptible to feedback inhibition have also been isolated and are used industrially to increase the yield of amino acids.

5

Food microbiology

5.1 How do microbes affect food?

An inevitable consequence of the start of agricultural practices that heralded the dawn of civilisation was the need to store crops from one season to the next. This brought with it the risk of microbial contamination and spoilage. The problems of **food spoilage** are, therefore, as old as civilisation. Not all spoilage, however, is detrimental. Fermentation, for example, can vastly improve the flavours of foods and can also turn an inedible food into a delicacy. The food processing industry has thus been exploiting microbes for centuries. Even so, it is only in the last 150 years or so that we have realised the contribution that microbes make to food production.

5.2 How are fungi used as food?

Probably the most familiar edible fungus is the mushroom. We have been eating mushrooms for centuries. The Romans valued mushrooms but were also very familiar with their potentially fatal effects. Sometimes they used this knowledge to great effect when someone important obstructed their ambitions. Agrippina murdered the emperor Claudius by feeding him mushrooms so her son Nero could succeed him as emperor. Nero then declared mushrooms the food of the gods, since it was mushrooms that had made Claudius a god. There is even a mushroom known as Caesar's mushroom (*Amanita caesarea*). This is a rather rare and highly prized specimen: a close relative of the death cap mushroom (*Amanita phalloides*).

Commercially available mushrooms in the UK have seen a recent and welcome expansion with a much greater variety of mushrooms available in the 1990s. Care must, however, be exercised when gathering wild fungi for eating. Most wild mushrooms are too bitter or tough or woody to make good eating, and although many wild fungi are edible, others are poisonous, even fatal. Only an expert may be able to distinguish edible mushrooms from their more dangerous relatives. The best advice is if you are in any doubt about the safety of a fungus that you are about to harvest, do not eat it. Rather, obtain your fungus from a reputable commercial outlet. It may not be as fresh as a wild mushroom, but at least it will be safe.

Commercial mushrooms (*Agaricus bisporus*) with their characteristic white caps covering dark brown gills are seeded on sterile straw or manure beds covered in a layer of soil. These are treated with the mycelia of laboratory-maintained stock. The soil and straw or manure are sterilised to prevent competition from undesirable contaminating species of fungus. Within a few weeks the mycelium will have spread throughout the bed and will then mature to produce the fruiting bodies or **sporophores**. As these mature they are harvested and sent for sale. A single seeding will produce a number of mushroom crops. The common commercial mushroom is a close relative of the field mushroom, *Agaricus campestris*. This grows during the summer and autumn in pastures and meadows and is especially abundant on ground that has been manured.

The oyster mushroom (*Pleurotus ostreatus*) is a yellow fungus now widely available in supermarkets. It makes an interesting alternative to the more common commercial mushroom. It has hardly any stem and in the wild grows as a bracket on decaying trees. Shiitake mushrooms (*Lentinus edulus*) are very popular in the Far East where they are commercially cultivated on hardwood logs. They are becoming very popular in the West, too.

Giant puffballs belong to the genus *Calvatia*. They are quite unmistakable and do not resemble other fungi. Providing they are eaten when young and while the inner tissues remain white they may be safely eaten. As they mature, spores develop within the puffball and they become powdery and inedible. Other edible puffballs belong to the genera *Lycoperdon* and *Scleroderma*. Puffballs make interesting alternatives to mushrooms. The spores of lycoperdon puffballs make lycopodium powder, beloved of physics classes.

Truffles are ascomycete fungi that grow as mycorrhizas. Traditionally pigs were used to hunt for truffles in woodlands. They find the smell an irresistible attractant. Dogs are now trained to sniff out truffles. They do not eat as much of the precious fungus as do pigs. The most highly regarded truffle is the black truffle, *Tuber melanosporum* also known as the Périgord truffle. It grows at the base of 'truffle' oaks and is used in the preparation of pâté de foie gras.

Some people consider the blewit (*Tricholoma saevum*), with its convex cap and wavy edges, among the best of the edible spring mushrooms. The fairy ring mushroom (*Marasmius oreades*) is also rated highly. Another spring mushroom is the common or true morel (*Morchella esculenta*). This is a mushroom than can appear from beige through dark olive to almost black and the cap is honeycombed. They are often found growing in apple orchards. Although the *New Larousse Gastronomique* says that there is little chance of confusing this with dangerous mushrooms, false morels belonging to the genus *Gyromitra* exist and occasionally cause illness in people who eat them. Again, the best advice is to urge caution if any doubt exists. The pore or boletus mushrooms may also cause confusion over identity. The edible mushrooms such as the king boletus (*Boletus edulis*) never have a trace of red in their stalk, unlike the dubious Satan's mushroom (*Boletus satana*). Despite its name this mushroom is only likely to cause severe indigestion.

Late summer and early autumn are seasons in which there is a rich harvest of fungi. Popular specimens include the columelle or parasol fungus (*Lepiota procera*) and the horse mushroom (*Psalliota arvensis*). Some members of the chanterelle family make particularly good eating. These are cup-shaped with vein-like gills and frilled edges. *Cantherellus cibarius* is the best example. The common name of its relative *Cantherellus cornucopioides* is the horn of plenty. This may, however, be a misnomer since its appearance is somewhat unattractive and its flesh is tough. Its smell is reminiscent of the truffle and it is for this reason that gastronomes enjoy it as a condiment. The French refer to this mushroom ominously as the trumpet of death.

There is a grey area when it comes to edible fungi. This is well illustrated by the shaggy inkcap mushroom (*Coprinus comatus*), which can be found from spring until autumn growing in lawns and gardens. Connoisseurs regard this as one of the choicest mushrooms, but great care must be taken when preparing this mushroom for eating that any blackened flesh is discarded. A further problem is that this may be confused with poisonous species that cause illness when eaten in a meal with alcohol. Indeed, consumption of alcohol up to five days after eating these mushrooms may precipitate illness.

In recent years there has been an increasing awareness of the importance of a healthy diet. This is an area where fashions come and go, and rigorous scientific evidence to support claims can be difficult to find. There can be no doubt, however, that many people would benefit from a healthier diet than that provided by the traditional British 'meat and two vegetables'. Fungi are relatively high in protein and contain very little saturated fat. Mycoprotein is the name given to artificial foods derived from fungi and they make a healthy alternative to meat in a balanced diet. The 1990s has seen the first successful application of biotechnology to provide a novel fungus food: Quorn. The starting material for the manufacture of Quorn is the wheat foot-rot fungus *Fusarium graminearum*. It is grown as a mycelial mat in a medium in which glucose is the main carbon and energy source: a cheap and plentiful starting material. The threads of mycelia lend the product a fibrous texture that we find very palatable. This also gives it an advantage over soya protein, which must be spun artificially into fibres before it can be used to make artificial meat substitute products. With the rise in vegetarianism, Quorn has made a successful start in the food market. It may well be the first of a number of mycoprotein products.

5.3   How are microbes involved in bread and alcohol production?

Bread is described as the staff of life. It is one of our most ancient processed foods and in one form or another most people still eat their daily bread. It is made by mixing yeast with flour and water in the process known as kneading. This distributes the yeast evenly through the dough. The mixture is allowed to 'prove', during which the yeast grows to produce large quantities of carbon dioxide. This becomes trapped as bubbles within the dough and these cause the dough to rise. The term **leavened bread** is used to describe bread made with yeasts. The bread mixture is kneaded for a second time to distribute evenly the carbonic acid produced by the yeast. Once the dough has risen again, bread is baked. A very hot oven may be used to start the baking to ensure that the yeast is killed, but the best bread is produced if the initial hot oven is cooled for the remainder of the baking period.

Baker's yeast, *Saccharomyces cerevisiae*, elaborates a variety of enzymes that enable it to break down the carbohydrates in flour starch to produce sufficient carbon dioxide to give leavened bread its characteristic texture. These include enzymes that release maltose and sucrose from starch as well as the respiratory complex necessary to produce the large quantities of carbon dioxide necessary for bread texture. Bread making is an aerobic process.

Baker's yeast does not simply cause the air holes to appear in bread. It also contributes subtle flavours. Just compare the taste of bread proved with bicarbonate of soda and cream of tartar with that of bread leavened by yeast. Other microbes may influence the flavour of bread. Sour dough is made by another yeast, *Saccharomyces exiguus*, used together with lactobacilli. These make an acid product that gives the bread its characteristic sour taste.

In contrast to baking bread, *Saccharomyces cerevisiae* can be exploited under anaerobic conditions. Without oxygen, yeast cells cannot respire. Rather, they rely on fermentation to provide their energy. The principal by-product of yeast fermentation is ethanol. Brewing is another very ancient human activity, as popular today as ever.

Wine making probably started by accident. When grapes and other fruits are harvested, they carry yeasts on their skins. Under the right conditions these can ferment the sugars within the fruit to produce alcohol. Today, winemaking is a major industry worldwide and the production of wine is carefully regulated. There is a staggering array of wines produced around the world, each with their peculiarities. Although much of the world's commercial wine is made from grapes, there has been a long tradition of making wine from other fruits as well. Winemaking that uses fruits other than the grape tends to be a domestic industry.

The starter material used in winemaking is known as the **must**. This consists of fruit suitably treated to make the fermentable substrates accessible to the yeast. With traditional winemaking this involved crushing grapes by trampling over them with bare feet in the process imaginatively known as treading the grapes. These days, grapes are more often mechanically pressed. White grapes will yield a white wine. Red wines are made from red grapes. Red grapes, can, however, yield white wines if the skins are removed before fermentation is established. Traditionally the yeasts on the grape skins, typically *Saccharomyces cerevisiae* and *Saccharomyces ellipsoideus,* fermented wine. This was a somewhat haphazard process and batches could easily become spoiled if 'wild' yeasts outgrew the fermentation strains early in the process. Starter cultures of the desired yeasts are now added, often after the must has been pasteurised or treated with sulphur dioxide to kill any microbes likely to spoil the finished product.

Fermentation is carried out without oxygen so that the yeasts produce alcohol. Wine typically contains 10–12% alcohol. The concentration of alcohol that accumulates during fermentation builds up to a point where it kills the yeast, preventing further fermentation. Fortified wines such as port and sherry are produced by a normal fermentation but they have their alcohol content increased by adding a spirit such as brandy once the wine fermentation is complete. In dry wines all the sugar in the original must is turned to alcohol. Sweet wines are made with musts that have a higher initial sugar content. Red wines have a heavier characteristic taste than white wines. This is because they contain tannins and other chemicals derived from the grape skins.

Fermentation produces a sediment of dead yeast cells. The fermenting wine is allowed to clear and is removed from the sediment in the process of

racking. In the best wines the clear product is aged in wooden casks. In some of the commercial vineyards where ageing would take too long, an artificial ageing is achieved by adding oak chips to the fermenting wine. Even in racked wines, sediments may accumulate over time in bottles, and older wines are often decanted before drinking. With sparking wines like champagne, the latter part of the fermentation process is carried out in the bottle. This leaves the problem of how to dispose of the sediment. During the fermentation the bottles are stored upside-down. The sediment is collected in the neck of the bottle. When fermentation is complete the necks of the bottles are frozen and the sediment is removed while the wine is held in place by the ice plug. The space left by removing the sediment is replaced with fermented wine and the bottles are then resealed.

Beers and lagers are made from cereal grains such as barley, wheat and even rice. The grains are first allowed to germinate to break down the starch stored in the seed. This allows the release of malt. Extraction of malt is continues when the germinated grain is mixed with water and is **mashed**. The mixture is heated gently to allow the extraction of soluble sugars and other material including peptones and amino acids released from the breakdown of proteins. These are essential for the growth of the yeast.

The longer the mashing process continues the darker will be the finished product. The clear liquid produced by mashing is known as **wort** and is separated from the solid residues. Hops are added to the wort at this stage. Hops are used to lend the beer its bitter flavour but they also release a variety of anti-microbial substances that probably act to prevent contamination of the beer while it is fermenting. The wort and hops are then boiled for several hours. This extracts the flavour from the hops and causes coagulation of any proteins remaining in the wort. It also sterilises the wort. The clear liquid is then run off the sediment, cooled and transferred to the fermentation vessel. Yeasts that remain at the bottom of the fermentation vessel produce lagers, hence they are known as bottom yeasts. Ales are brewed using top yeasts. These float to form a scum over the top of the fermentation vessel. At one time it was thought that a different species of yeast was used to produce lagers from that used to brew ales. Both drinks are now considered to be the result of fermentation by *Saccharomyces cerevisiae*, although lager yeasts will probably be known as *Saccharomyces carlsbergensis* for some time to come. Ale fermentation is carried out at a higher temperature than lager brewing: 14–23°C as opposed to 6–12°C. Consequently ales take only five to seven days to produce while lagers take 8–14 days. Lagers are also stored at a lower temperature than ale.

The production of many spirits follows the same basic process as beer making but the alcoholic content of the drink is considerably increased in the subsequent distillation process. The distillate contains not just alcohol but other volatile products that give the spirits their characteristic flavours. In the case of gin made from the distillation of grain ferments, juniper berries are added to give the drink its characteristic flavour. Malt distillates yield whisky (Scotch) or whiskey (Irish). Sugar cane is used to make rum. Vodka is made from potatoes. The characteristics of the product depend on the manner and length of the ageing process. A good ten-year-old Scotch can lose up to 10% of its volume during the maturation process, said to be the share belonging to the angels. Distillers believe that they take this in exchange for the secret of distillation.

5.4 How are fermented vegetables and meats produced?

Fermentation is one of the principal methods whereby microbes obtain energy. The most familiar end-product of fermentation is alcohol but different fermentative pathways produce a variety of products. Some may have desirable attributes. Microbial fermentation of foods is an ancient process and is practised in every society around the world. It must originally have arisen as a random, accidental process but is now highly regulated. There are considerable advantages in fermenting foods. Products of fermentation act as preservatives, considerably increasing the shelf life of a fermented product. They also change the flavour of the food and can considerably enhance the taste of the product. Microbial metabolism may also increase the vitamin content of fermented food and may break down chemicals in the food that are otherwise indigestible by humans. This increases the nutritional value of the fermented product. Despite these advantages, fermentation is essentially a process of controlled microbial food spoilage but producing a desirable rather than a distasteful product. Spoilage is, after all, just a question of taste.

5.4.1 Sauerkraut

Sauerkraut is a cabbage product considered to be a delicacy in Germany, where it is often eaten with ham. It is prepared by the fermentation of cabbages: a process that reduces slightly the nutritional quality of the cabbage. Fermentation does, however, make the food more digestible. People who complain that it is indigestible often ignore the rich foods such as the smoked meat they eat with their sauerkraut. Rich food is much more likely to cause indigestion than the sauerkraut that accompanies it. Firstly the cabbages are washed in an attempt to remove any unwanted contaminating microbes that may be found on their outer leaves. This washing will leave the natural microflora largely undisturbed. The cabbage heads are then shredded and soaked in brine with an optimum salt concentration of between 1.8 and 2.25%. Layers of cabbage leaves can be interspersed with layers of salt and juniper berries. These are used to give added flavour to the sauerkraut. The salt causes carbohydrates to leach out of the cabbage shreds and these sugars act as substrates for fermentation by a succession of bacteria. Initially, species such as *Leuconostoc mesenteroides* and *Lactobacillus brevis* ferment sugars to provide a variety of organic products such as lactic acid, acetic acid, ethanol and mannitol. Because these bacteria produce a variety of fermentation products but principally lactate, they are known as **heterofermentative bacteria**. The pH of the fermentation drops to about pH 3.5 as a result of this phase of production. After this, fermentation is taken over by the **homofermentative** *Lactobacillus plantarum*, which produces only lactic acid. This second phase of fermentation uses up mannitol and this removes the bitter taste from the maturing sauerkraut. It also further reduces the pH of the product to about pH 2. If the salt concentration in the brine is higher than optimal, the secondary fermentation is not carried out by *Lactobacillus plantarum*. Rather, it is undertaken by heterofermentative bacteria such as *Enterococcus faecalis* and *Pediococcus cerevisiae*. At salt concentrations above 3%, a very tough product is produced. This is known as 'pink kraut'. The colour of this product is the result of overgrowth by carotenoid-producing yeasts of the genus *Rhodotorula*.

5.4.2 Dill pickles

Dill pickles are fermented cucumbers and not just the butt of Martian jokes. The green pickle often found in commercial burgers from fast food outlets is a dill. Huge brine vats are used to contain the fermenting fruits. The brine leaches out the sugars from the cucumbers and these are fermented by bacteria from the normal flora found on the surface of the fruit. Streptococci start the fermentation and as the pH falls *Leuconostoc* and *Pediococcus* species and *Lactobacillus plantarum* continue the process. Carbon dioxide is generated as a by-product of fermentation. To prevent this accumulating and causing bloat damage, fermentation vessels are purged with nitrogen to flush out the carbon dioxide. Natural fermentation may take a few days to begin and to assist this process acetic acid may be added to the brine at the start of the fermentation.

This not only accelerates the initiation of fermentation but it also helps to prevent the growth of spoilage organisms by lowering the pH of the initial reaction below the optimum for growth of spoilage organisms.

5.4.3   Other fermented vegetable products

There are a number of foods where the involvement of microbial fermentations is not obvious. Olives are rendered edible only after fermentation with *Lactobacillus plantarum* and *Lactobacillus mesenteroides*. Fermentation is also important in the production of coffee and chocolate. The tough outer coats of coffee beans are removed after treatment with a variety of bacteria. These include *Erwinia dissolvens, Leuconostoc* and *Lactobacillus* species and yeasts of the genus *Saccharomyces*. These do not impart additional flavour to the coffee bean. In contrast, the microbes that assist in the removal of the coats from cacao beans do confer a characteristic taste to cocoa and chocolate. Soy sauce, used as a seasoning for Chinese food, is made from a mixture of soya beans and rice fermented by a variety of bacteria and fungi. These include *Lactobacillus delbrueckii, Aspergillus oryzae, Aspergillus soyae* and *Saccharomyces rouxii*.

5.4.4   Fermentation of meats

Meat products such as salami and bologna sausages are produced by fermentation with *Pediococcus cerevisiae, Lactobacillus plantarum* and halophilic members of the genus *Bacillus*. Heterofermentative bacteria produce a variety of organic compounds as a result of their fermentation reactions. Such microbes are always employed in the manufacture of these meats because they generate a diversity of acid products. These not only act as preservatives but also confer a characteristic subtle flavour on the sausage. In contrast to bacterial fermentations exploited in the production of salami, country cured ham is produced following fermentation with fungi of the genus *Aspergillus* and the genus *Penicillium*. Izushi is a Japanese delicacy made from a mixture of fish, rice and other vegetables and is produced by fermentation with lactobacilli.

5.4.5   Silage production

Not only humans benefit from fermented foods. Farm animals are fed on silage during the winter months when fresh food is in short supply. Trips into the countryside will often take the visitor past large silos, used to produce silage. Silage is prepared by the fermentation of grass, kale, sugar beet heads, chopped corn and similar vegetable material. Young, succulent crops make much better silage than do mature plants. Production of silage depends upon strictly anaerobic fermentation of the vegetable matter. Exposure to air permits the overgrowth of aerobic spoilage organisms. The starting material is tightly packed to exclude air. Molasses, the syrup drained from raw sugar, may be added to assist the initial fermentation. During the early stages of silage fermentation, bacteria such as members of the family Enterobacteriaceae and *Aerobacter* species predominate but as the pH of the silage falls, streptococci and lactobacilli rapidly become the dominant bacteria. After fermentation for about three or four weeks, a stable product results. Fermentation makes a sweet smelling product that is almost irresistible to livestock. The smell may not be so acceptable to humans more used to an urban environment.

The acids formed during silage production are corrosive and in older silos these may damage the fermentation container. This, in turn, may permit air to gain access to the fermenting silage, leading to spoilage of the product. To overcome this problem, silage production today is often carried out in stacks tightly enclosed in polythene. During the late summer months, the British countryside is dotted with armies of large black plastic sacks packed with vegetable matter destined to become silage.

5.4.6   Fermented dairy products

Many humans lack the ability to break down or absorb lactose. This sugar is, therefore, able to enter the lower gut where it is metabolised by the commensal flora to produce acid and gas. This leads to **flatulence** and diarrhoea. Because of this, many people worldwide do not drink fresh milk. Rather, they prefer to consume fermented milk products in which the lactose has been metabolised to lactic acid. Fermentation of lactose in the milk rather than in the gut converts lactose to lactic acid. This can then be eaten safely in the knowledge that there will be no antisocial or painful consequences such as those caused by lactose intolerance.

The most familiar fermented liquid milk product in the Western world is yoghurt. Milk is first heated to kill its indigenous flora and is then mixed with a starter culture comprising a 1:1 mixture of *Lactococcus thermophilus* and *Lactobacillus bulgaricus*. During fermentation, acid is formed primarily by the lactococcus, whereas the aroma largely results from the activity of the lactobacillus. The lactic acid produced in yoghurt considerably lowers its pH. This inhibits the growth of spoilage organisms. Consequently, yoghurt has a reasonable shelf life, particularly compared with the shelf life of unfermented milk. Yoghurt has also been adopted as a folk remedy for vaginal thrush even though its efficacy has never been established in proper trials. One reason for its application is the mistaken belief that the lactobacilli in yoghurt will supplement those of the commensal flora. The lactobacilli found associated with humans are of a different type.

In parts of the USSR and in Scandinavia, kefir is a very popular drink made by the fermentation of mares' milk. Elsewhere camel's milk is the starting material for this product. It is a fizzy, alcoholic drink in which the fermentative microbes occur in 'grains'. Many different microbes have been found in kefir, including yeasts, lactobacilli, lactococci and leuconostocs. There is a considerable regional variation in kefir because in different geographical locations the precise mixture of microbes responsible for fermentation will vary. Because of this, a variety of fermentation products will be generated, allowing differences in the finished drink. The drink is also somewhat modified by the length of the fermentation process. After a day's fermentation, mild kefir is produced. This has slight laxative properties. Medium kefir is made by fermentation for two days. The laxative properties of mild kefir are not found with medium kefir. Mature kefir is fermented for three days. It can be slightly constipating if taken in excess. Kumiss is a similar alcoholic fermented milk. Lactose intolerance is a problem that affects the majority of humans. Our response is reflected in the prevalence and diversity of alcoholic fermented milk drinks that are available across the world.

Acidophilus milk has become a very popular health product made by the fermentation of milk with *Lactobacillus acidophilus*. Its popularity can, in part, be attributed to the belief that *Lactobacillus acidophilus* exerts a beneficial effect on the commensal flora of the lower gut. It has not, however, been established that bacteria in the milk can survive passage through the stomach and then go on to colonise the bowel. Because of this the health benefits of acidophilus milk remain theoretical.

Butter has a very ancient and respectable pedigree. It probably originated with nomadic tribes where it was used to prolong the storage of milk. Abraham in the Old Testament is said to have used butter as a 'symbolic food', offering it as a precious gift. It can be made from the milk of cows, sheep, goats, mares, asses or camels, each yielding a characteristic odour and taste. This is attributable to bacteria such as *Lactobacillus diacetylactis* found in the milk used in its manufacture. These help to give flavour and odour to butter.

Imagine being the first person to eat cheese. Your milk had not only gone off and turned sour but it had also probably solidified. It may even have acquired a mouldy crust. Perhaps the first person to eat cheese was very hungry indeed. The reward, however, must have been enormous, as reflected in the astronomical numbers and diversity of cheeses that are now available. The art of cheese-making is very ancient. Originally the processes involved were random but now cheese making is highly regulated. Cheeses are highly varied in nature but all are fermented milk products and the manufacture of every cheese follows the same basic pattern.

The milks from cows, sheep and goats may all be used in the production of cheese and the nature of the starting material fundamentally affects the finished product. This has been recognised for at least 300 years. Daniel Defoe, describing Cheddar cheese production, wrote 'Every meal's milk makes one cheese and no more, so that the cheese is bigger, or less, as the cows yield more, or less, milk. By this method, the goodness of the cheese is preserved and, without all dispute, it is the best cheese that England affords, if not the whole world affords'. Perhaps it is not surprising that Defoe did not admire that other great English cheese, Stilton. His description of it reads 'Stilton . . . cheese, which is called our English Parmesan and is brought to table with mites, or maggots round it, so thick that they bring a spoon with them for you to eat the mites with, as you do the cheese'. These days, cheese making is a far more hygienic process.

Cheeses may be broadly classified into two groups, unripened and ripened cheeses. Unripened cheeses like cottage cheese, cream cheese and Mozzarella are soft and are made by the lactic acid fermentation of milk. These are blander in flavour than the ripened cheeses, which undergo secondary processing following the lactic fermentation. Many bacteria are employed in the lactic fermentation of milk to produce cheese. Amongst the most common are *Lactococcus lactis*, *Lactococcus cremoris* and *Leuconostoc cremoris*. Characteristic combinations of bacteria produce characteristic cheeses. Originally, the presence of these bacteria was a matter of chance and influenced by local strains. Today, most milks are pasteurised before fermentation and specially prepared starter cultures are added. This is to establish a better-controlled production process. When single strains are used as starter cultures, the resultant cheeses tend to be bland in flavour. Fuller flavoured cheeses are produced from mixed starter cultures.

A significant problem in cheese production is that the atmosphere inside dairies can build up significant levels of bacteriophage and these can devastate bacteria used in starter cultures. This is a particular problem in dairies that employ only a very limited number of strains in their starter cultures.

Another problem met with in modern cheese production is the presence of antibiotic residues in the milk taken from herds treated for mastitis. The antibiotics may be present in sufficient concentrations to inhibit starter cultures. Coagulase-negative staphylococci that produce penicillinase are sometimes added with the starter culture. The penicillinase produced by these bacteria breaks down the antibiotic, and the lipases that are also produced may enhance the flavour of the final cheese. In this way a virtue may be made of a necessity.

Starter cultures produce flavour changes from the milk and produce acid. This causes the formation of curds as a result of the coagulation of milk proteins. This process may be enhanced by the addition of rennin, an enzyme once only obtainable from calf stomachs but now also manufactured by genetically engineered microbes. The solid curds are heated, then pressed to separate the whey, or liquid milk products. The pressed curds are then salted and may be subjected to various ripening processes.

Ripened cheeses come in various degrees of hardness. They range from the soft ripened cheeses like Brie and Camembert and semi-soft cheeses such as Roquefort and Stilton, through the hard cheese like Cheddar to very hard cheeses like Parmesan. The hardness of the final cheese reflects the length of the ripening processes. Soft cheeses are ripened for between one and five months. Hard cheeses take from three months to a year to ripen and very hard cheeses may be ripened from a year to 18 months. Ripening may also be associated with the secondary action of bacteria or fungi.

The blue veins of cheeses such as Stilton and Roquefort are caused by the growth of *Penicillium roquefortii*. This is now deliberately added after curding but originally it was present as an aerial contaminant of the creameries and caves in which these cheeses were made. Even now the environment in which they are produced influences the crusts that develop on such cheeses. Stilton cheese supports moulds, yeasts and bacteria on their surface. The precise microflora on the rind can be used to identify the individual creamery in which it was made. The surfaces of Camembert and Brie cheeses are inoculated with *Penicillium camembertii* and this develops into the skin on these cheeses. Limburger cheese is soaked in brine to encourage the growth of *Brevibacterium linens*. The same bacterium can be isolated from people who suffer from smelly feet. The holes or 'eyes' found in Swiss cheeses such as Emmental are the result of gas production by *Propionibacterium shermanii*, cultures of which are now added before curding. Not all cheeses are ripened in the presence of microorganisms. Edam cheese is dipped in wax before ripening to prevent access of microbes during the maturation process.

5.5 What role do microbes have in food spoilage and preservation?

It may be thought that food should be prepared and eaten in as fresh a condition as possible. There are, however, exceptions to this. Game birds, hares and venison are hung to mature before they are considered fit to eat. This alters the texture and the flavour of the meat, but not everyone finds this type of food palatable. With the organisation of modern society, consumption of fresh food cannot always be achieved. Perhaps this was always so. It has been argued that the beginnings of agriculture brought in the dawn of civilisation. This involved the gathering and storage of food for future use, rather than relying on continual foraging. Consequently, humans have always lived with the problems of food storage and preservation. Today foodstuffs still have to be stored, sometimes for considerable periods before being sold. They may also be transported over very long distances, even between continents. Even after purchase they may undergo another prolonged period of storage before consumption, with or without cooking. During all this time food is potentially liable to microbial spoilage. A number of food preservation techniques have evolved to prevent the spoilage of food.

5.5.1 How do microbes cause food spoilage?

Food spoilage is the process whereby food is rendered unfit to eat. This may be because of an unpalatable taste, odour or texture. Physical damage may render food unpalatable because of its appearance and spoilage may also result from chemical reactions. An example of chemical spoilage is rancid butter, caused by the oxidation of its components. Food spoilage may have a biological origin, either as a result of intrinsic enzymatic activity in the food or from the activity of other organisms. Spoilage results particularly, although not exclusively, from the action of bacteria and fungi. Foods vary considerably in their susceptibility to spoilage. Meats and dairy products are highly susceptible to spoilage and are considered perishable. Other foods, including fruits and vegetables, are considered semi-perishable or non-perishable. In many instances, spoilage results in the production of unacceptable flavours, odours or textures in the food that is spoiled. **Food poisoning** is a special case of food spoilage. In food poisoning, the incriminated food typically appears perfectly wholesome and yet it contains poisonous chemicals, toxins or live microorganisms that can cause illness and even death.

Meat and dairy products provide humans with an excellent source of proteins, fats and carbohydrates, all in forms that can be easily exploited. Unfortunately, these also act as nutrients for a vast array of spoilage organisms such as the pseudomonads. These may have a variety of undesirable effects. Spoilage organisms are **proteolytic**, breaking down proteins to release amino acids, amines, hydrogen sulphide and ammonia. Hydrogen sulphide is often known as rotten egg gas with just cause. The proteins in egg have a high sulphur content. Fats are made rancid by their breakdown to glycerol and free fatty acids. Carbohydrates are fermented to produce acids, alcohols and gasses. Joints of meat only have microbes over their surfaces and the total microbial load is relatively small. This means that joints are less prone to spoilage than minced meat and sausages. The manipulation necessary in the production of these items contributes significantly to the spread of contaminating microorganisms. The surface area of these foods is also considerably increased as a result of processing. This greatly increases the surface area over which microbes can grow. This increases the microbial load on the food. Consequently, the risks of spoilage of mince and sausages are far higher than that of the Sunday joint.

The reducing agents present in meat provide an environment in which putrefactive anaerobic bacteria can multiply even when the meat is exposed to the air. Paradoxically obligate anaerobes grow less well in fine mince. This is because of its greatly increased surface area exposed to oxygen in the atmosphere. The temperature of meat storage has a profound effect upon the type of spoilage organism present. When meat is held at room temperature, Enterobacteriaceae are common spoilage bacteria, along with staphylococci, micrococci and aerobic Gram-positive sporing bacilli. Refrigeration suppresses these bacteria but permits the growth of other spoilage organisms, particularly pseudomonads. Besides bacteria, moulds can also cause spoilage of meat. Black spots on meat are caused by the growth of *Cladosporium* species. White spots are caused by *Sporotrichum carnis* and yellow or green spoilage patches are caused by *Penicillium* species. The rainbow effect seen on ageing refrigerated bacon is because iridescent spoilage bacteria grow on the meat. Even more spectacular can be the spoilage organisms seen sometimes on fish. Photobacteria are aquatic microbes that can break down ATP to produce visible light. After a couple of days, photobacteria may grow sufficiently to enable raw fish to glow in the dark.

Milk is an excellent growth medium for bacteria, moulds and yeasts. Consequently it may undergo spoilage in different ways as a result of the activity of various microbes. Souring of milk is the result of acid fermentation by lactobacilli. The principal product of fermentation is lactic acid. In turn, moulds and yeasts may break this down. Proteolytic bacteria of the genus

Bacillus cause the degradation of milk proteins. Eventually this may lead to the formation of a clear straw-coloured liquid product. Growth of capsulated organisms such as *Lactococcus cremoris* and *Enterobacter aerogenes* causes the formation of 'ropy' milk. The milk does not become separated into curds and whey but it comprises rope-like strands in a liquid matrix and is very unpleasant.

Members of the genus *Clostridium* can attack milk to produce different effects. Some, such as *Clostridium sporogenes*, break down milk proteins to produce alkaline products. Other clostridia, such as *Clostridium septicum,* produce acid from the fermentation of lactose. Acid or alkali production may easily be shown by the addition of litmus to the milk. *Clostridium sordelii* is a clostridium that causes milk proteins to clot and *Clostridium novyi* forms gas from milk. The most spectacular reaction with milk, however, is the formation of a stormy clot caused by *Clostridium perfringens.* This bacterium produces acid from the fermentation of lactose and causes the milk proteins to form a clot. So much gas is produced that it rips through the clot, giving it a characteristic stormy appearance. The spoilage of milk by clostridia is a very useful tool in the identification of members of this genus.

Fruit and vegetables do not contain as much protein as do meats and dairy products. Typically, moulds that can penetrate the outer layers of the fruit or vegetable initiate spoilage. Penetration breaches a primary defence against infection. This damage permits access of other microorganisms to otherwise inaccessible tissues. Secondary invaders can then go on to inflict further damage to the produce. Bacteria such as those of the genus *Erwinia* that cause soft rots typically cause secondary infection of fruits and vegetables. There are stories of potato stores rapidly being turned into a black ooze as a result of the degradative properties of an erwinia infection.

5.5.2   How can food be preserved?

Food is derived directly or indirectly from living organisms, which all undergo natural decay. Similarly, all foods will undergo decay. This may result from the activity of **autolytic enzymes** intrinsically present in food, or from the activity of contaminating microbes. Contaminating microorganisms on or in food may be derived from the commensal flora of the animal or plant from which the food is derived, or may result from handling during food processing. Many of these organisms are harmless but some cause spoilage and others may produce food poisoning.

Handling and processing of food increase its microbial load. This increases the potential for its spoilage and for food poisoning. Great efforts are made by the food industry to make safe the food we eat. **Aseptic handling** of food is important in minimising risks in food establishments and practices are tightly regulated by the force of legislation. Control of spoilage and the preservation of food is under constant review and new control techniques are being developed, but food preservation practices extend back into pre-history.

Cooking of food

Humans have cooked food since time immemorial. The cooking process has various benefits. It causes the denaturation of intrinsic autolytic enzymes, tenderises food, imparts desirable flavours and odours to the food and it alters the microbial flora, eliminating many heat-sensitive organisms. Cooking, however, does not always kill microorganisms. Even vegetative bacteria can survive certain cooking regimens. If this were not so, then the incidence of salmonella infections would be very significantly reduced, for example. Furthermore, heat-resistant spores survive many cooking regimens. Prolonging cooking times or raising cooking temperatures in an attempt to produce a sterile product often results in an unacceptable finished dish. There may be a conflict between a desirable food and one that is microbiologically safe. The soft-boiled egg affords a good example of cooking that produces a gastronomically tempting food that is not necessarily safe. What would you dip your soldiers in if eggs were held at a temperature necessary to kill salmonella?

Overcooking may produce an unacceptable food. Milk is easily rendered unpalatable by heat exposure. Raw milk is also a vector for many serious and potentially fatal infections. To render milk safe for consumption, pasteurisation regimens have been developed. Originally this was to control spread of tuberculosis and salmonella infections. A 'low temperature holding' (LTH) process was used. This, however, was not sufficient to kill *Coxiella burnetii*, the cause of Q fever. Consequently a 'high temperature, short time' (HTST) process was developed. Pasteurisation does not indefinitely increase the shelf life of the milk because milk naturally carries thermotolerant microbes. A further significant problem is the post-pasteurisation contamination of milk. This has considerable implications for the storage of pasteurised milk. Ultra-heat treatment (UHT) of milk yields a product that does have a significantly increased shelf life, measured in months rather than days. Milk is exposed for a very short time to temperatures about 130–135°C. This process, however, does not irreversibly inactivate enzymes in the milk. In time, these break down the milk fats to produce 'off' flavours. Heat-resistant proteases also add to the degradation of UHT milk. Perhaps this is why some people joke about ultra horribly tasting milk.

Canning and bottling provide a useful extension to the cooking process. Cooked foods are **hermetically sealed** in metal cans or in bottles. In commercial canning, the cooking process is designed to reduce to an absolute minimum the risk of *Clostridium botulinum* spores surviving. Other more heat-resistant spores may survive the canning process and subsequently cause spoilage of the canned food. *Clostridium botulinum* is the cause of botulism, discussed in Section 5.6.6. It is the most serious food-borne illness. Spoilage may also occur if the seal of the can or bottle is faulty, permitting the entry of microbes. Such a failure was responsible for 515 cases of typhoid that occurred around Aberdeen in 1964. The source of this incident was traced to tinned corned beef imported from Argentina. In this case, contamination of the food did not produce a detectable change in its appearance. In other instances, gasses evolved during the microbial fermentation of the food cause swelling of the can, giving it a 'blown' appearance. This typically happens when a yeast contaminates canned fruit. Blown cans may also result from the chemical reaction between acids in the food and the metal of the can causing the production of hydrogen. Lacquer applied to the inner surface of cans overcomes this problem. If, however, a metal can becomes dented, the lacquer may become damaged and chemically blown cans may result. People who eat from damaged cans do so at their own risk.

Low-temperature preservation of foods

For centuries food has been kept at low temperatures to help to delay spoilage; in the case of wines and cheeses, low-temperature storage also assists in the maturation and ripening processes. Caves maintain a constant low temperature and they have long been used to store food, particularly cheeses and wines. Large country houses frequently had icehouses. These are large structures, often built underground for extra insulation. Ice from local rivers and the estate lake was collected during the winter months and stored in the icehouse. Because of the size of the ice stock and the insulation of the structure, the ice gathered one year would last through the summer and into the next winter. This provided ice for summer desserts. The ice also provided a cool environment and food was often stored in these buildings to delay its decomposition.

The modern domestic refrigerators are the humble descendent of the icehouses of the great estates and they have revolutionised food storage. Refrigeration temperatures in domestic appliances are typically 1–10°C. Unfortunately, most refrigerators in the home operate at the upper end of this range. Holding at temperatures just above freezing significantly increases the shelf life of food. It slows the growth of many spoilage organisms and most food-poisoning organisms. It also retards the activity of intrinsic autolytic enzymes, although it is not without problems.

Infections with *Listeria monocytogenes* were once rare but in recent years have become much more common. Food acts as the vector for infection. *Listeria monocytogenes* grows very well at the temperatures found in most domestic refrigerators. It will even grow slowly at 0°C. Very large numbers of listeria may accumulate during prolonged storage of food in refrigerators. This problem has been considerably exacerbated by the appearance of 'cook-chill' foods. These are meals that are sold pre-cooked and stored under refrigeration but at temperatures above freezing. These convenience foods may be eaten cold directly after purchase, but some cook-chill foods require warming before consumption. Often microwave ovens are used to reheat cook-chill meals and these frequently do not heat the food to a sufficient temperature to kill any listeria present. The problem is made worse by the fact that *Listeria monocytogenes* is relatively heat tolerant. Besides posing a microbiological threat, prolonged storage of cook-chill food results in a significant loss of nutrients and vitamins as a result of autolytic processes in the food.

Freezing of foods overcomes many of the problems of refrigeration above the freezing point of water. Many foods are not, however, suitable candidates for freezing. The commercial freezing of delicate foods like strawberries involves flash freezing in liquid nitrogen. This effects an extremely rapid transition to the frozen state. Water molecules do not have time to form large ice crystals, which could damage the structural integrity of the food. This process is very expensive and, therefore, more resilient foods are blast-frozen. There is a particular problem with the freezing of fruits and vegetables. Intrinsic autolytic enzymes continue to work during storage and when the food is thawed it becomes limp. This effect may be minimised in some cases by **blanching** the vegetables by immersing them in boiling water for a very short time just before freezing.

Freezing may result in the loss of viability of a microbial culture but it cannot be relied upon to kill microorganisms. Indeed, the infectivity of viruses may increase after freezing. Typically, freezing causes up to 50% loss in the viability of vegetative bacteria. Spores, however, because of their low water content, can be frozen and thawed repeatedly without adverse effects upon their viability. Bacterial culture collections may even be stored in a cryoprotective glycerol buffer for many years at temperatures of about −70°C. Water does not freeze completely inside cells until temperatures below −50°C. At higher temperatures cellular processes continue, albeit at a very slow rate.

Thawed food may decompose faster than it would had it not been frozen. This is more frequently a result of ice damage and the action of autolytic enzymes rather than the effects of the microbial flora of the food. The spoilage organisms of thawed frozen foods may differ from those found on fresh produce. Frozen peas are more likely to be contaminated with streptococci and with bacteria of the genus *Leuconostoc* derived from the processing plants than are fresh peas. These bacteria produce large quantities of slime as well as 'off' flavours not associated with fresh peas.

Physical methods of food preservation

Several physical methods are employed in the preservation of food. These include cooking, chilling, drying, osmotic protection, filtration and irradiation. Some processes are new but others are of much more ancient pedigree. In the case of dried food, an ancient technique underwent a radical adaptation to meet the needs of space travel. This brings food technology into the twentieth century and provides us with such essentials as instant coffee and tea, dried soups and potted rice or noodle snacks.

Food preservation by drying

Drying of food has been used since the dawn of civilisation. Most of the water is removed from food and its associated microbial flora. Intracellular constituents in dried food are no longer in solution and the removal of water prevents metabolic processes that rely on reactants being dissolved together. Dried fruits and pulses are very common in the Western world and in the East dried meat and fish are very popular. Not all dried foods are dehydrated to the same extent. Dried fruits often retain some water and these may be subject to spoilage by moulds that can grow with the very little water that is available to them.

Although drying of food is a very ancient practice, it has recently been coupled with freezing in the process of **lyophilisation**. Food is first flash frozen and is then dried under vacuum. Lyophilised foods weigh little because of the removal of all their water. Freeze-dried food simply requires reconstitution with water and heating, if appropriate, before they are eaten. Dried milk and soup powders and instant coffee are familiar examples of food prepared by freeze-drying but other foods may be prepared in this way. Complete meals may be lyophilised for use by astronauts and explorers going to remote areas and for whom weight limitation is an important consideration.

Osmotic protection of food

Dehydration of food has the effect of reducing the water available to cells, inhibiting their metabolic processes. The same effect may also be achieved by placing foods in hypertonic conditions so that water is lost from cells by the process of osmosis. This results in **plasmolysis** of microbes. There are two principal processes that exploit osmosis to preserve food: salting or the generation of a high sugar environment.

Salting was used to preserve food in the ancient civilisations of Egypt and Rome. Even at moderate salt concentrations, food-poisoning organisms are destroyed. Most spoilage bacteria and yeasts cannot tolerate salt concentrations of about 10%. Despite this, moulds can still grow on salted foods and can cause spoilage. High sugar solutions also exert an osmotic protective effect. This property has been exploited for centuries in the manufacture of jams, marmalades and other preserves. The sugar concentration in foods has to be considerably higher than salt and commonly a minimum sugar content more than 50% is used. Despite this, it is thought that part of the preservative properties of condensed milk is conferred by the moderately high concentrations of lactose in this food. As with salted foods, moulds can cause spoilage of sweet preserves exposed to the air. Just try leaving an opened bottle of jam in a cupboard for a few weeks and you will be amazed at the moulds that you will acquire as pets.

Filtration

Liquid foodstuffs can successfully be preserved by filter sterilisation. Filtration is used in the production of wine, beer, cider, fruit juices, maple syrup and soft drinks. Strictly filtration does not sterilise the product, since viruses can pass through the smallest of filter pores.

Irradiation of food

Considerable excitement heralded the introduction of irradiation of food as a means of preservation. It is a process that does not involve heating of food and cobalt-60 used in irradiation is a cheap by-product of the nuclear industry. Food irradiation works by converting water to hydrogen peroxide and by generating free radicals that react with cellular components, causing damage. In practice, in many situations this has unfortunate side effects. It may produce unacceptable changes in the irradiated food, causing tainted odours and unpleasant tastes. This is particularly true where higher doses of irradiation are required to make foodstuffs safe. This considerably limits the useful application of food irradiation. It has, however, found a limited use in the treatment of spices and has been employed to prolong the shelf life of soft fruits such as strawberries. Low-dose irradiation has been proposed to assist the elimination of salmonellas from poultry and to suppress the spoilage flora found on meats, poultry and fish, thereby extending the shelf life of these foods. There is considerable public opposition to the irradiation of food. This, in part, is because of the mistaken belief that irradiated food itself becomes radioactive. It does not. Public hostility, however, coupled with the limited range of foods that are suitable for irradiation will probably restrict developments in this area.

Modified atmospheric packaging of foods

Packing food under vacuum in plastics that exclude oxygen has been found to prolong considerably its shelf life. Such packaging material has allowed the deliberate manipulation of the atmosphere under which the food is held. A wide variety of foods are now stored under modified atmospheres in the process of **modified atmospheric packaging** (MAP). These include fresh raw meats and poultry, cooked meats, cheese, pasta, salads and various sandwiches. The ratios of atmospheric gasses are modified when the products are first packed to suppress the spoilage flora present in or on the food. Frequently this is achieved by elevating the level of carbon dioxide and reducing the amount of oxygen present in the modified atmosphere. It should be remembered, however, that as a result of the metabolic processes of the indigenous flora of the food, modified atmospheres within packages undergo considerable changes over time. In container-packaged foods being transported over long distances, the atmosphere can be continually monitored and modified within the container to retain constant conditions. This is referred to as **controlled atmospheric packaging** (CAP).

Vacuum-packaged meats are now very common and represent another class of modified atmosphere. The absence of oxygen prevents the growth of the aerobic flora on the meat. Increased levels of carbon dioxide that develop under vacuum-packaging enhance the antimicrobial effect. Carbon dioxide is particularly inhibitory to pseudomonads.

Chemical preservation of food

Many chemicals have an antimicrobial activity. Some methods of chemically preserving food are very ancient: others are more recent in origin. Some foods are intrinsically antimicrobial. Spices such as sage and rosemary, together with garlic and onions, contain large quantities of essential oils that exert a considerable antimicrobial effect. This may be a beneficial side effect of their culinary use to enhance the flavour of the cooked food. Fermentation of foods helps to preserve them as well as endowing desirable gastronomic qualities upon them. Vinegar is a natural product used to pickle foods such as onions that do not have a suitable flora for natural fermentations. The pH of the pickling vinegar is too low to support the growth of most spoilage organisms, at least in the medium term. The name vinegar is derived from the French word *vinaigre*, meaning sour wine. It may be produced from a variety of starting materials including grapes for wine vinegar, apples for cider vinegar and barley for malt vinegar. The starting materials are responsible for the different flavours of the resulting vinegars but fundamental to all these products is the production of ethanol prior to the formation of acetic acid. Yeasts ferment extracts of the starting material to produce a product containing about 10% ethanol. This product is then repeatedly trickled over beds of bacteria of the genus *Acetobacter*. Traditionally, these beds were supported by birch twigs or beech wood shavings. Today a variety of supports, organic and inorganic, can be found. An abundant supply of air remains essential for the success of this production process. The acetobacters oxidise the ethanol present to produce acetic acid. The product of this dual process is vinegar, used to the same effect in artificial pickling as are the acid products in naturally fermented foods.

Smoking of fish and meat is an ancient practice found in both the Old World and New World civilisations. The process involves drying of the exposed food surface and this itself inhibits the indigenous microbial flora. Furthermore, smoking causes the deposition of aromatic compounds in the flesh that is being cured. These include formaldehyde, phenolics and cresols. These compounds have considerable antimicrobial properties. Some fish are cured in a process known as hot smoking. The temperature of the fish is raised to 70°C for 30 minutes. This further reduces the microbial load on the food and enhances its shelf life, as well as imparting a pleasant flavour. Smoke houses can gain an enormous international reputation for the excellence of their fish; one example is the Arbroath Smokie, but there are others that can produce fish of superb quality, such as the smoke house found at the foot of the cliff close to the ruins of Whitby Abbey in Yorkshire, which produces excellent fish.

More recently, artificial preservatives have been added to foods. Sodium benzoate is a chemical widely used to preserve food. Sulphur dioxide and sodium metabisulphite are added to home-brewed beer and wine and also to some of the cheap commercial wines. Salts of sorbic and propionic acid are added to help to preserve bread, and nitrates and nitrites are added to meats and meat products. The use of these chemical preservatives is controversial, particularly in the case of nitrites. These compounds are potent mutagens and there is a theoretical risk that prolonged consumption of nitrites may lead to an increased risk of developing cancer. This has not yet been confirmed by observation.

5.6   What causes food poisoning?

Almost every food that we eat is a potential vector for food poisoning, for a variety of reasons. It may result from chemical contamination, the nature of the animal or plant tissue consumed or the presence of microorganisms or their by-products. Unlike food that has been subjected to the activity of spoilage organisms, food associated with food-poisoning incidents usually appears to be perfectly wholesome and therein lies the danger. It is only through constant vigilance and a high standard of domestic practice that humans avoid food poisoning.

The investigation of food poisoning involves examination of different types of epidemic incident. Many individuals may be affected by food from a single source. Family food poisonings occur when some or all the members of a single family exhibit symptoms of food poisoning. Sporadic cases involve individuals only. Epidemic incidents may be explosive if a large number of people in a single institution or attending the same function are simultaneously affected. Alternatively, cases may be dispersed if food from a point source is transported over a wide area. The easiest incidents to study are epidemics and the most difficult to examine are sporadic incidents. Because of the high incidence of sporadic cases, it is very difficult to assess the true scale and nature of food poisoning. This problem is exacerbated by the fact that in many cases food poisonings are a self-limiting condition of relatively short duration. Consequently, such cases are never reported to the official bodies responsible for keeping records. As a result, the true incidence of food poisoning may be between 10 to 100 times greater than official figures suggest. Of all the causes of food poisoning, the most common is bacterial food poisoning.

5.6.1   Chemical contamination of food

Chemical contamination of food may have a natural or an artificial cause. Heavy metal ions may be naturally present in the soil or may be found there as a result of industrial contamination. Plant tissues can concentrate heavy metal ions very efficiently when growing in soils that either naturally or as a result of contamination contain high levels of these ions. Chronic heavy metal poisoning may result from consumption of contaminated vegetables over a period. The soils around Somerset, England may have very high levels of cadmium, and fungi isolated from this vicinity may contain up to 30% dry weight cadmium. People are advised not to eat garden produce from affected areas because of the greatly increased risk of heavy metal ion poisoning.

It has been suggested that the Ancient Romans suffered from lead poisoning because of their use of lead piping. There is, however, no archaeological evidence to support this claim. Furthermore, most Roman drinking water was supplied in terracotta pipes and not lead ones. Ironically, their love of fast-food style restaurants could well have been a source of bacterial food poisoning for the Ancient Romans. Apparently they did frequently suffer food-poisoning incidents.

Artificial chemical poisoning of foods may result from modern agricultural practices. Pesticides and herbicides are now widely employed, together with an array of feed additives for livestock. This is all designed to improve the yield obtained from the land. Furthermore, much processed food contains preservatives to increase the product shelf life. Many preservatives are chemicals to which we would not otherwise be exposed. The long-term cumulative effect of ingesting such compounds together with our food has yet to be adequately established and the subject remains controversial. Chlorinated diphenyl insecticides such as DDT were very popular when they were first introduced. There is now good evidence that these compounds are highly toxic not only to insects but also to higher animals, including humans. Symptoms of DDT poisoning include dizziness, nausea, headache, muscular tremors and weakness. These symptoms may progress to paralysis, coma and ultimately may cause death. Because of this, the use of chlorinated diphenyls is now banned in many parts of the world, except in very special circumstances.

5.6.2 Food poisoning associated with consumption of animal tissues

Fish meat may act as the direct cause of food poisoning and other meats may act as vectors for protozoal, tapeworm (cestode), fluke (trematode), or roundworm (nematode) infection. All of these have serious consequences for human health.

Puffer fish and the Moray eel have flesh that may prove fatal to humans if eaten. Deaths from this type of food poisoning are more common in the Far East than in Western countries. This reflects the popularity of sushi dishes in Japan. Even in the West, eating fish can cause problems. Fish of the sub-order *Scombroidei* include mackerel (*Scomber scombrus*) and tuna fish (*Thunnus thynnus*). The flesh of these fish is rich in tryptophan. When such fish are inadequately stored, the tryptophan breaks down to produce histamine. When contaminated fish are eaten, symptoms of scombrotoxin poisoning result. Ingestion of the affected fish causes an allergic-type reaction and symptoms include a burning sensation in and around the mouth, nausea and a transient skin rash.

Filter-feeding shellfish can concentrate plankton as a result of their feeding habits. Certain dinoflagellate plankton elaborate a toxin that may be concentrated by filter-feeders like mussels and oysters. When humans subsequently eat these shellfish, cases of paralytic shellfish poisoning may result. The duration of this condition is one to three days and is associated with a significant mortality.

The obligate intracellular protozoan *Toxoplasma gondii* forms cysts that are quite often found in pork and lamb and less frequently in beef. Many people worldwide have an asymptomatic infection caused by *Toxoplasma gondii*, but occasionally a clinical condition, toxoplasmosis, results from the ingestion of infectious cysts. Toxoplasmosis is associated with swollen and painful lymph glands, malaise and chronic fatigue, similar to the symptoms of glandular fever caused by the Epstein–Barr virus.

The tapeworm *Taenia sagginata* causes infection following consumption of raw or undercooked beef. Cooking temperatures of 56°C for as little as five minutes is sufficient to prevent infection but this temperature must be maintained at the centre of the joint to be effective. Salting of beef has a similar beneficial effect. Different species of tapeworm are associated with other meats. *Taenia solium,* for example, is associated with the consumption of undercooked pork.

The nematode worm *Trichinella spiralis* causes trichinosis, a diarrhoeal disease linked with inflammation and muscular weakness in humans. Pork is the usual source of human trichinosis. Recently cases in the USA have been associated with the consumption of bear meat, which is said to be very tough and not particularly tasty. These appear to be three good reasons for not eating bear meat. Nematodes of the genus *Anisakis* are associated with fish. Consumption of raw or undercooked fish has resulted in human infection. This is more common in Japan than in the West but the rising popularity of sushi in Europe and North America is reflected in an increase of cases in these areas.

The liver fluke *Fasciola hepatica* has a complex lifecycle involving both mammals and a freshwater snail. Human cases of liver fluke infection arise from the consumption of watercress harvested from beds where infected snails live.

5.6.3   Food poisoning associated with the consumption of plant material

Many plants or their component parts may be poisonous if eaten. Occasionally plant extracts that are potentially lethal have been usefully harnessed in the service of clinical medicine. Foxgloves (*Digitalis purpurea*) provide a source of the heart drug digitalis and deadly nightshade (*Atropa belladonna*) is a source of atropine, another heart drug. Women used extracts of deadly nightshade as eye drops to dilate their pupils. This was done to make them appear more beautiful: hence '*belladonna*', beautiful lady. It may still be used as a pupil dilator in medical investigations. Even plants that are widely used for culinary purposes may have parts that are poisonous. For example, the stems of the culinary rhubarb plant (*Rheum palmatum*) are used in stewing, jams and in tarts and pies. Rhubarb roots were a traditional purgative and the leaves of the rhubarb plant are rich in oxalic acid. This causes irritation of the mouth and the oesophagus. Ingestion of rhubarb leaves induces vomiting and abdominal pains.

Improper preparation of vegetable material may also prove to be dangerous. Overlong soaking of dried pulses, together with inadequate cooking before consumption pre-disposes the consumer to symptoms of food poisoning resulting from toxins found in the pulses.

5.6.4   What are food-borne infections?

Food may be a vector of life-threatening infections such as typhoid, cholera and dysentery and for less dangerous infections such as those caused by the thermophilic campylobacters. Bacteria that spread by the faecal–oral route and are often associated with drinking water that has been contaminated with human or animal faeces cause these infections. Diseases such as cholera and typhoid are principally diseases of the Third World as these diseases have largely been eradicated in developed countries by the introduction of adequate sanitation programmes. In most developed countries, these diseases are generally found only in travellers returning from areas where the diseases are endemic. Even when travellers take great care not to drink untreated water while on holiday, they are often not quite so scrupulous about the food they eat, particularly fruit and salads. Eating foods that have been washed in contaminated water may result in a life-threatening infection.

Some microbiologists draw an important distinction between infections of this type, referred to as food-borne infections and other bacterial food poisonings. In food-borne infections, the food merely acts as a vector of the infection. The infectious agent, such as *Salmonella typhi, Vibrio cholerae* or *Shigella dysenteriae* does not multiply on or within the contaminated food. In other food poisonings, the causative agent is not only capable of multiplication within the implicated food but probably requires to do so to attain a sufficiently high infectious or toxic dose.

Food-borne infection caused by thermophilic campylobacters

One of the most significant food-borne infections in developed countries is caused by the thermophilic campylobacters. Although recognised for many years as animal pathogens, it was not until 1977 that the role of thermophilic campylobacters as a major cause of human disease was realised. They are now found to be the most common cause of acute infective **enteritis** in Northern Europe and North America and the incidence of campylobacter infections is rising.

Symptoms of campylobacter enteritis comprise a profuse and watery or slimy diarrhoea, sometimes with blood present in the stools. A cramping abdominal pain, nausea, fever and, occasionally, vomiting accompany the diarrhoea. Campylobacters produce a toxin that is genetically related to the cholera toxin, although its effects are less severe. Consumption of contaminated food and unpasteurised milk has been implicated as a cause of campylobacter enteritis and so too has drinking water that has suffered faecal contamination. Campylobacters do not replicate in food and so campylobacter enteritis is considered a food-borne infection rather than a food poisoning. The incubation period of campylobacter enteritis varies considerably, from about 15 hours to several days. The length of the incubation period is dependent upon the infectious dose taken by the victim.

Four groups of thermophilic campylobacters cause human enteritis: *Campylobacter jejuni*, *Campylobacter coli*, *Campylobacter lari* (once named *Campylobacter laridis*) and *Campylobacter upsaliensis*. The relative frequency of isolation of campylobacters depends on the geographical location of infection. In Canada and the UK, *Campylobacter jejuni* accounts for 98% of human campylobacter isolates. In Poland and parts of Central Africa, however, up to 40% of campylobacters isolated from humans are identified as *Campylobacter coli*. Campylobacters also show an animal host preference. Most campylobacters isolated from cattle are identified as *Campylobacter jejuni*, whereas most pig isolates are *Campylobacter coli*.

Poultry is a major source of human infection caused by campylobacters, as is drinking unpasteurised milk. Pork is an infrequent cause of campylobacter infections in humans. Other foods associated with human campylobacter infections include turkey and raw or undercooked fish, particularly filter-feeding shellfish. In an unusual incident, mushrooms were implicated in campylobacter infection. In this case, however, the ultimate source proved to be horse manure used to grow the mushrooms rather than the fungi themselves.

Chickens are the source of about half of the human cases of campylobacter infection and are a particular hazard if eaten raw or undercooked. In many incidents, however, the source of infection cannot be traced. Even when the source can be identified by epidemiological methods, it is not always possible to confirm such findings by microbial culture. This problem is compounded by the fact that campylobacters easily revert to a viable but non-cultivable state. These may return to a cultivable form only after passage through a susceptible animal host.

Wild birds act as an unusual vector of campylobacter infection. Many species of bird act as hosts to a variety of campylobacter species. Magpies (*Pica pica*) and jackdaws (*Corvus monedula*) have been observed pecking the tops from milk bottles left on doorsteps during domestic milk deliveries. As a result of this feeding adaptation, the milk becomes contaminated by campylobacters, even if it had previously been pasteurised. The contaminated milk may then act as source of human infection.

5.6.5 What is bacterial food poisoning?

Bacterial food poisoning is of two principal types, intoxication or infection. In practice there may be an overlap between these two classes. In bacterial intoxication, a toxin is produced and is responsible for eliciting the clinical manifestations of the disease. It may not be necessary to ingest viable bacteria to suffer from an intoxication-type food poisoning. In contrast, in food poisoning resulting from bacterial infection, viable bacteria must be ingested and must replicate within the gut before symptoms become apparent.

The incubation period of the illness is often a reflection of the nature of a particular food poisoning incident. For example, after ingestion of the toxin, *Staphylococcus aureus* poisoning is apparent from 30 minutes to six hours later. In contrast, the average incubation period for salmonella enteritis is between 15 and 48 hours, depending upon the infective dose. This longer incubation gives sufficient time for bacterial replication to occur in the gut before the appearance of symptoms.

5.6.6 What is bacterial intoxication?

Staphylococcus aureus *food poisoning*
In the UK *Staphylococcus aureus* is possibly the most common cause of food poisoning but the actual number of cases is very difficult to determine. The reason for this is that the disease is of very limited duration, with symptoms persisting for less than half a day. The principal symptoms of staphylococcal

food poisoning are vomiting and prostration. These occur from one to six hours after eating the contaminated food. Diarrhoea is uncommon in staphylococcal intoxication. Food poisoning strains of *Staphylococcus aureus* produce a heat-stable enterotoxin that has a direct effect upon the central nervous system.

The ultimate source of *Staphylococcus aureus* causing food poisoning is the commensal flora of a food handler. Many healthy carriers of *Staphylococcus aureus* exist and people with septic spots are more likely to be carriers. Spread of staphylococci to food may be facilitated if the food handler implicated indulges in unsociable behaviour such as nose picking and bum scratching. Lax methods of food preparation and storage contribute to the problem of staphylococcal food poisoning.

Many foods harbour small numbers of *Staphylococcus aureus* and these may be eaten safely. If foods are not properly stored before consumption, however, staphylococci may flourish and produce sufficient exotoxin to cause disease. Chicken and other meat sandwiches that have been left at room temperature and are becoming stale are an important cause of staphylococcal intoxication. Sandwiches that have curled edges are prime candidates to act as vectors of staphylococcal intoxication. Other foods commonly associated with *Staphylococcus aureus* include cream cakes and custards, along with other dairy produce. Foods that require manual manipulation also pose a threat. For example, prawn cocktails contain dairy produce as well as the prawns. Dairy products are recognised vectors of staphylococcal food poisoning and a further problem comes from the prawns. Their exoskeletons must be removed by hand. This delicate operation increases the risk of contaminating the shellfish meat considerably. Cold cooked meats, especially ham, are also an important source of *Staphylococcus aureus* food poisoning. Staphylococci are salt tolerant and can multiply without competition from other bacteria that cannot tolerate the salt in hams.

Bacillus cereus *food poisoning*

Bacillus cereus was at one time associated with a clinical condition indistinguishable from *Staphylococcus aureus* food poisoning and referred to as 'Chinese restaurant syndrome'. Happily, since the cause of this condition has been elucidated, the problem is now largely of historical importance. *Bacillus cereus* is a spore-forming aerobic Gram-positive rod often found contaminating rice. The spores survive the boiling necessary to cook rice. It used to be common practice to refrigerate any unused boiled rice at the end of one day so it could be used as a source of fried rice on the following day. Often this involved placing a large bulk of rice into a refrigerator and allowing it to cool slowly.

As the rice cools in this way, spores may germinate and bacteria multiply. Because of the bulk of the rice, cooling may be very slow, allowing a long time for bacterial multiplication and toxin formation. The emetic toxin produced by *Bacillus cereus* can survive the rapid cooking employed in the production of fried rice. It is very heat stable and can withstand temperatures of 126°C for more than an hour and a half. In contrast some strains of *Bacillus cereus* produce an enterotoxin that is heat labile and is quickly destroyed at temperatures above 56°C. This toxin causes a diarrhoeal disease with a longer incubation period. Once eaten, the emetic toxin can cause disease from half an hour to six hours later. As with *Staphylococcus aureus* food poisoning, Chinese restaurant syndrome caused by *Bacillus cereus* is a self-limiting disease of short duration. Another Chinese restaurant syndrome exists and this is associated with a sensitivity to monosodium glutamate. The symptoms manifest as a hot flush and constitute a chemical food poisoning.

Clostridium perfringens *food poisoning*

Clostridium perfringens is an obligately anaerobic Gram-positive bacillus that produces spores. In artificial media, spores do not readily form but in the gut *Clostridium perfringens* sporulates with ease. It may be found in small numbers as part of the resident commensal flora of the bowel and on the skin of the lower limbs. It is capable of growing at temperatures up to 49°C, and at 44°C the mean doubling time is only 12 minutes. With this rate of growth, it takes only four hours for a single bacterium to yield over one million progeny. Although *Clostridium perfringens* is an obligate anaerobe, it can tolerate traces of oxygen. These features make *Clostridium perfringens* a bacterium well suited to cause food poisoning.

Symptoms of food poisoning are caused by an enterotoxin and typically comprise a mild diarrhoea together with a colicky abdominal pain. Vomiting is not a feature of *Clostridium perfringens* food poisoning, although nausea may be a symptom. Food poisoning caused by *Clostridium perfringens* is often mild and does not usually result in constitutional upset. The illness usually resolves within a single day. Fatalities are very rare and are often only seen in elderly, frail patients, for whom dehydration is a major problem. Unusually for a bacterial food intoxication, symptoms do not typically occur until 12–24 hours after eating the contaminated food. This is because the bacteria multiply and sporulate within the gut and the enterotoxin is produced as a secondary metabolite *in situ*.

Clostridium perfringens food poisoning incidents occur most often in institutions such as hospitals, schools or university halls of residence, where meals are prepared in large quantities. The incriminated food is often a bulk-prepared meat or poultry stew, soup or meat and potato pie that has been cooked on one day and warmed through and eaten on the following day. Spores survive the initial cooking but other bacteria do not survive. This removes the competition for growth. *Clostridium perfringens* flourishes without the inhibitory effects of the flora present on uncooked meat. Because of the large volumes of food involved, cooling is often very slow, permitting adequate growth of clostridia to occur to provide a sufficiently high infectious dose. The many reducing agents present in meats help to mop up the residual oxygen, which in turn helps to support the growth of the oxygen-tolerant anaerobe *Clostridium perfringens.*

A much more serious but very rare form of *Clostridium perfringens* enteritis is caused by strains producing the β-toxin. A family in Norwich, Norfolk fell victims of this condition. They prepared a steak and kidney pie on a Sunday but did not eat it until the next day. Because they did not have a refrigerator the pie cooled only slowly and was held at ambient temperature overnight. This provided ideal conditions for the multiplication of *Clostridium perfringens.* All four members of the family died as a result of eating the contaminated steak and kidney pie.

Clostridium botulinum *food poisoning*

The first description of botulism occurred over 200 years ago and was associated with the consumption of contaminated sausages. *Botulus* is the Latin word for sausage. *Clostridium botulinum* is widely distributed in nature. The species is divided in eight serovars, A to H, based upon the antigenic structure of the toxin produced. Serovars A, B, E and occasionally F are implicated in human disease. Serovar E is often found in marine environments. Type C toxin causes disease in fowl and serovar D is implicated in cases of bovine botulism. The optimum temperature for toxin production is 30°C but type E toxin can be produced at temperatures as low as 6°C. The bacteria that produce this toxin can grow at a temperature of 3.5°C. This reflects its marine habitat and association with fish.

Toxins from *Clostridium botulinum* serovars A, B and E have been purified. They are heat-labile neurotoxic peptides with similar molecular masses. It is estimated that the lethal dose of botulinum toxin for humans is only about 1–2 micrograms and hence it is one of the most poisonous of all natural products. Botulinum toxin is absorbed from the gut and is carried to the nervous system where it causes a flaccid paralysis, in which the muscles are relaxed. The toxin acts by preventing production or release of the neurotransmitter acetylcholine in nerve synapses, thus interfering with the communications between nerve cells.

Symptoms of botulism occur within a day of consumption of the contaminated food. The disease manifests itself in several ways. Visual disturbances result from the inability to coordinate the eye muscles and patients often suffer from double vision because of this. Inability to swallow is caused by paralysis of the oesophageal muscles. Paralysis of the muscles of the throat prevents speech. The disease progresses rapidly over a few days. Mortality among people with botulism is very high and death results from paralysis of the respiratory muscles or from cardiac arrest. Patients remain fully conscious until just before death.

Since spores of *Clostridium botulinum* are widely distributed in nature, they may easily be found contaminating vegetables, fruit and other foodstuffs. Botulism is very rare in the UK, largely because of strict controls placed upon the food processing industry. It is more common in the USA where home canning of produce is much more popular. Botulism is associated particularly with the consumption of home-canned food that does not require reheating. Heating food to 80°C for only one minute inactivates botulinum toxin and simmering food for 20 minutes before consumption renders contaminated food safe to eat.

Spores of toxigenic strains of *Clostridium botulinum* present on foodstuffs may survive inadequate food processing and may enter defective cans after processing is complete. This may be as a result of cross-contamination by raw food, or because of unhygienic practices. This is exemplified by the tinned salmon incident described below. Spores will germinate and thrive in the anaerobic environment provided by the canned or vacuum-packed foods, particularly since potential competitive organisms will have been destroyed in the processing. Toxic foods may become spoiled by some strains of *Clostridium botulinum* and these foods are clearly unfit for consumption. Equally, many toxigenic strains of *Clostridium botulinum* do not cause apparent food spoilage and these are the strains that are responsible for botulism.

Since 1922 there have only been nine reported cases of botulism in the UK. Botulism most often occurs as family cases. A recent and unusual incident in the UK involved 27 people who were affected after eating hazelnut yoghurt. The illness was not as severe as usual. This was because the acidity of the yoghurt limited growth of the bacterium. In this incident, only one fatality was recorded: an old lady who died after eating the incriminated food. When the ultimate source of the toxin was discovered it proved to be the canned hazelnut purée. This is an excellent example of the potential problems posed by large-scale canning operations. The previous British incident, occurring in 1978, is a more typical case of botulism. It involved two couples who ate a tin of salmon as a special treat for Sunday tea. By 2.00 a.m. on the Monday morning, the first couple were suffering breathing difficulties. The second couple were traced and all four were admitted to hospital. Despite antitoxin therapy, two of the four victims died, one after 17 days, the other after 23 days. The source of the bacterium was traced to a canning factory in Alaska. Workers were in the habit of drying their wet contaminated overalls on the cans that had been processed and were left out to cool, thus making use of a free source of heat. The implicated can had been flawed and the bacteria gained entry with subsequently fatal consequences.

5.6.7   What food poisoning is associated with bacterial infection?

Salmonella infections

Salmonella typhi causes typhoid, also known as enteric fever. This is a disease that is mainly contracted through drinking contaminated water but it may be caught after eating fruit and raw vegetables. *Salmonella typhi* can be easily distinguished from other salmonellas by its inability to produce gas from glucose and on the basis of its amino acid metabolism.

Salmonella cholerae-suis is another salmonella that can cause serious infection. It is a bacterium that is often associated with pigs and it has a tendency to be more invasive than other non-typhoid salmonellas. Consequently, it causes more extraintestinal infections such as osteomyelitis, meningitis and pneumonia.

The salmonellas associated with food poisoning are sub-divided into more than 1800 varieties based upon the possession of different combinations of O and H antigens (Fig. 5.1). The classification is based upon the scheme devised by Kauffman and White in the 1930s and confusion in using this scheme may arise since each serovar is given an apparent species name. The development of newer taxonomic techniques has shown the folly of assigning a species name to strains that may only differ from one another by the presence or absence of sugar molecules occurring in the lipopolysaccharide component of the outer membrane.

It is still a common convention to refer to serovars as though they were individual species. While acknowledging the problems of defining a bacterial species, particularly concerning taxonomy based on antigenic structure, we have adopted the practice of using serovar names as if they were individual species. There is still a debate regarding the taxonomy of this genus and one of the most important issues centres upon changing the names of very familiar organisms. Most clinicians have no difficulty in recognising that for a

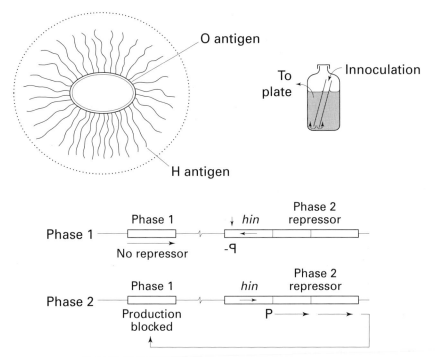

Fig. 5.1.   Salmonella antigens. The O antigen is associated with lipopolysaccharide in the bacterial outer membrane while the H antigen is associated with flagella. These are responsible for motility and salmonella cells can be either motile or non-motile. In addition to the O and H antigens, *Salmonella typhi* has a capsular antigen, shown as a dotted line in this figure. It confers virulence and hence is known as the Vi antigen. To isolate salmonella in the motile phase, a Craigie tube can be used. The culture is inoculated through a hollow tube in the semi-solid medium. Only motile bacteria can swim out to be recovered outside the tube. Phase variation in salmonella is controlled by a genetic switch. The *hin* region encodes a recombination enzyme that causes the DNA to 'flip' orientation. It also contains the promoter that regulates expression of two genes: one responsible for production of the phase 2 (non-motile) flagellin, the other a repressor protein that prevents expression of phase 1 flagellin. When the switch is in one orientation, the phase 2 flagellin and repressor are produced. When the switch is inverted, these proteins can no longer be produced and so the phase 1 flagellin can be produced instead.

human the isolation of *Salmonella typhi* has much graver consequences than the isolation of *Salmonella typhimurium*. If these familiar names were to change as a result of taxonomic pedantry, then the ease with which the different reports are now assessed would be lost, with potentially fatal outcomes.

Salmonellas exist in two phases, being either motile or non-motile. Depending upon the phase of the bacterium, different H (flagella associated)

are expressed. This is controlled by a genetic switch where a section
A that encodes the flagellar antigen is associated with a site-specific
ination system and becomes inverted. This 'flip-flop' causes the gene
ng the phase 2 antigen to be expressed when the DNA is in one orientation but not when the sequence is inverted. Full identification of a serovar
requires that the H antigens in both phases should be examined. In practice
this is only done in reference laboratories.

Although there are many serovars of salmonella, very few are common
causes of human disease. In the UK, *Salmonella typhimurium* accounted for
about half of all the human infections caused by salmonellas from the pre-
war years until the mid 1980s. During the 1980s, however, one phage-type of
Salmonella enteritidis, phage-type 4, became increasingly prevalent and, at the
time of writing, this single phage-type is responsible for more salmonella
infections of humans than all the other salmonellas together. This serovar is
commonly associated with chickens. The principal reason for the rise in its
prevalence is that this strain, like others such as *Salmonella gallinarum* and
Salmonella pullorum, has evolved the ability to infect the egg-laying apparatus of
hens. As a consequence they can infect eggs. The temperature of the yolk of
a soft-boiled egg ranges from 37 to 42°C, ideally suited for the growth of salmonellas rather than for their destruction. Because the temperature of soft-
boiled egg yolks is insufficient to kill salmonellas, many *Salmonella enteritidis*
infections are associated with raw or undercooked eggs or egg products, such
as home-made ice cream and fresh mayonnaise. It should be remembered,
however, that very few eggs are contaminated in comparison with the total
number of eggs sold each day and so the risk to the individual consumer is
slight. It has been estimated that an individual would have to eat one soft-
boiled egg a day for 40 years to become infected with *Salmonella enteritidis.*

The rise to prominence of *Salmonella enteritidis* is not confined to the UK.
Evidence published in post-revolutionary Russia show that during the 1980s
the Soviet Union was suffering a dramatic rise in *Salmonella enteritidis* infections.
These were also associated with consumption of eggs or egg products and the
timing of the event was the same as in the UK. Such information was not
forthcoming under the old political regime. Eggs have also caused problems
associated with salmonellas in the past. In the period before the Second World
War, very few serovars of salmonella were found in the UK. Large quantities
of powdered eggs were imported from China to supplement our home-
produced supplies during the Second World War. The egg powder was not
pasteurised and it acted as a vector for a diversity of serovars, many of which
persist today. Now, all egg products for commercial use must be pasteurised.

Salmonellas are widely distributed in nature and have been isolated from many animal species including poultry, ducks, cattle, pigs, domestic pets and reptiles. During the early 1970s over 300 000 cases of human salmonella infection were caused by contact with turtles kept as domestic pets. This became a serious problem during the craze for the Mutant Ninja Turtles films, when children bought large numbers of these animals for pets. Other reptiles also carry salmonellas. People who keep iguanas, for example, may acquire salmonella infection from their exotic companions, who carry the bacterium as a commensal organism.

Intensive farming methods have contributed to the relentless increase in salmonella infections seen in recent years. Animal protein is now processed and re-cycled as animal foodstuffs for use with intensively reared farm animals. This food supply is frequently contaminated with salmonellas, which may colonise or infect the animals eating these feeds. The cramped conditions under which some animals are kept also provide ideal conditions for cross-contamination and cross-infection.

There is further ample opportunity for the cross-infection of animal carcasses and offal at slaughter. At least 50% of the poultry sold for human consumption carry salmonellas. Freezing does not kill salmonella and if frozen meat and poultry is not adequately thawed before cooking it presents a considerable risk of salmonella food poisoning. Raw meat and poultry should also be prevented from coming into contact with other foods that may be consumed without further cooking, otherwise these may become contaminated with salmonellas from the raw meat or poultry. Consumption of unpasteurised milk is another important source of salmonella infections in humans.

In the USA in the early 1980s, 85 people became infected with *Salmonella muenchen*. The distribution of cases spread from California to Massachusetts. Epidemiological studies linked this with the illicit use of marijuana. The supplier had, presumably, tried to increase his profit margin by diluting the drug with animal faeces and this was the probable ultimate source of the *Salmonella muenchen*. This incident is important because it established that a substance may act as a vector for salmonella infection even if it is not eaten. In this incident, the rolling and smoking of joints was responsible for many infections. Faecal contamination was also the cause of an unusual incident in which bean shoots were the vectors of infection by *Salmonella saint-paul*.

The incubation period for salmonella infections is typically from 15 to 48 hours. Symptoms range from an inapparent infection, through typical gastrointestinal problems, to life-threatening Gram-negative septicaemic shock, with some symptoms resembling those of typhoid. Septicaemic shock is characterised by shaking chills, rapid heart beat, low blood pressure and mental confusion. The most common manifestations of a salmonella infection are a moderate diarrhoea together with some vomiting and an associated abdominal pain. The typical infectious dose of salmonella for a fit young adult is of the order of 10^6 to 10^7 bacteria, but under special circumstances this may be significantly reduced. Chocolate has been implicated as a vector of infection with both *Salmonella napoli* and *Salmonella eastbourne*. In each incident, the infectious dose was of the order of 100 organisms. Chocolate is thought to exert a protective effect upon the bacteria as they pass through the acid environment within the stomach. Also, the observation that these incidents involved children was probably also significant. Children and the elderly are more susceptible to salmonella infection.

Antibiotic therapy should be avoided unless salmonella infections show evidence of spreading outside the gastrointestinal tract. In the most vulnerable of patients such as the newborn, the elderly or the immunocompromised, antibiotic therapy should be considered, despite the usual advice. Antibiotics do not significantly shorten the clinical course of the disease. They also prolong the period of asymptomatic carriage and hence the shedding of salmonellas in faeces. In the UK about 3% of the adult population at any one time are asymptomatic carriers of salmonellas. This figure is higher among food-handlers, who are more likely than the general population to come into regular contact with salmonellas. Consequently, food-handlers are more likely to contract salmonella infections. Food-handlers who are also salmonella carriers do pose a threat to the public health and major food-poisoning incidents have arisen as a consequence. The risk should be minimal, however, providing the personal hygiene of the carrier is scrupulous.

Escherichia coli *infections*
Certain serovars of *Escherichia coli* have been associated with diarrhoeal disease in very young children or in travellers. *Escherichia coli* diarrhoea is either of a type that resembles mild cholera, with a watery diarrhoea, or is similar to dysentery, with severe abdominal pain and mucosal damage as well as diarrhoea. A heat-labile toxin that is genetically related to the cholera and campylobacter toxins causes the first type: the second results from the action of a heat-stable toxin similar to the Shiga toxin of *Shigella dysenteriae*. Although food poisoning caused by *Escherichia coli* is most often associated with raw or undercooked meat, one of the largest incidents to occur to date was caused by consumption of contaminated Brie cheese. Several hundred people were affected in this incident.

Recently *Escherichia coli* O157 has been linked with a severe condition known as haemolytic-uraemic syndrome, in which a diarrhoeal episode precedes acute haemolytic anaemia and renal failure. Anaemia and renal failure are caused by the action of circulating toxin rather than as a direct consequence of infection and may occur a considerable time after initial infection. The disease is most often associated with consumption of inadequately cooked beef, particularly large, thick beefburgers. The disease is associated with the production of a verotoxin, so called because it causes a cytopathic effect in cultured cells called Vero cells. This condition has a high mortality. *Escherichia coli* O157 can be easily detected in stool samples through its inability to ferment sorbitol. Substituting sorbitol for lactose in MacConkey's medium can make an indicator medium for these strains. This bacterium has recently come to prominence because of a cluster of infections occurring in Scotland resulting in a number of fatalities. During the waterlogged Glastonbury Rock Festival of 1997, nine people were infected with *Escherichia coli* O157. In this incident it is thought that they acquired the infection from the ubiquitous mud, contaminated with cattle dung.

Yersinia enterocolitica *infections*

Yersinia enterocolitica is a Gram-negative facultative bacterium. It is a non-lactose fermenting member of the family Enterobacteriaceae frequently isolated from pigs. It has been linked with a human gastrointestinal infection that clinically mimics appendicitis. The bacteria are invasive and may be isolated from the mesenteric lymph nodes removed from infected patients undergoing surgery for appendicitis and where the appearance of the appendix is normal. It is capable of growth at low temperature, such as is found in the domestic refrigerator. In this respect it resembles the Gram-positive *Listeria monocytogenes*.

Listeria monocytogenes *infections*

Listeria monocytogenes is a Gram-positive bacillus with a very wide natural distribution. It is capable of growing at temperatures below 0°C and holding of samples at 4°C is used as an enrichment method for *Listeria monocytogenes* since it will outgrow other bacteria under such conditions. Furthermore *Listeria monocytogenes* is relatively heat resistant. After changing pasteurisation conditions from low-temperature holding (LTH) method to high temperature for a short time (HTST), *Listeria monocytogenes* has occasionally been isolated in pasteurised products. Low-temperature holding involved treatment at 62.8°C for 30 minutes, whereas high-temperature, short-time pasteurisation involves exposure at 71.7°C for just 15 seconds. It is not certain whether the presence of *Listeria monocytogenes* in these foods results from bacteria surviving the pasteurisation process or whether post-pasteurisation contamination occurs with this very common environmental bacterium.

Another feature of *Listeria monocytogenes* that enhances its role as a food-poisoning organism is its ability to survive on hands for long periods. Experimentally contaminated fingers support *Listeria monocytogenes* for at least eight hours and bacteria are not easily removed by conventional hand washing. Therefore, handling of infected food can inoculate hands, which may then go on to contaminate other foods: this is a good example of cross-contamination. Foods associated with *Listeria monocytogenes* include dairy products, especially soft cheeses, pâtés and vegetable dishes such as coleslaw. The increased popularity of cook-chill food has also contributed to a dramatic rise in the incidence of *Listeria monocytogenes* infections. Cook-chill food is pre-prepared and then refrigerated at temperatures above 0°C. This is ideal for the multiplication and enrichment of *Listeria monocytogenes*. Much publicity was generated as a result of the increased incidence of listeriosis; according to certain commentators this amounted to 'listeria hysteria'. One beneficial effect of this was that following the intense public interest, and hence education, the number of cases fell sharply. Unfortunately, as people have become complacent the trend in listeria infections has again turned upwards.

The infective dose of *Listeria monocytogenes* is not known. Many people regularly consume small numbers of *Listeria monocytogenes* without adverse effect, but pregnant women and the immunocompromised are at particular risk. *Listeria monocytogenes* causes meningitis and/or septicaemia in newborn babies and people who are immunocompromised. It also causes premature labour, still birth and septic abortions when consumed by pregnant women. Because of the risk of listeria infections, pregnant women are advised not to eat soft cheeses and pâtés.

Bacillus cereus *infections*

Besides causing a food intoxication, *Bacillus cereus* may cause a long-incubation food poisoning. The average time of incubation is similar to that of *Clostridium perfringens* food poisoning, with which it is often clinically confused. The symptoms include diarrhoea and cramping abdominal pains. This illness is caused by a different enterotoxin from that associated with rice intoxication. The toxin involved in rice intoxication is heat stable, whereas this toxin is heat labile. The illness generally lasts less than half a day. Vomiting is not a feature that commonly occurs in this form of *Bacillus cereus* food poisoning.

Vibrio parahaemolyticus *infections*

Although very rare in the UK, almost half of all the food-poisoning cases in Japan are caused by *Vibrio parahaemolyticus*. This is a marine bacterium found in the warm offshore waters around Japan, America and southwest England.

It requires a relatively salty environment and almost all food-poisoning cases have been attributed to eating raw or undercooked fish, including shellfish. The relative prevalence of this bacterium as a cause of food poisoning in Japan is a reflection of the national diet of that nation compared with a typical Western diet.

The only way to prevent *Vibrio parahaemolyticus* food poisoning is to cook fish thoroughly. The incubation time is from 6 to 48 hours, depending upon the infecting dose of bacteria and how much pre-formed enterotoxin is present in the food. Symptoms include diarrhoea and vomiting, often with moderate or severe abdominal cramps. The enterotoxin of *Vibrio parahaemolyticus* has not been fully characterised but it does cause loss of body fluid into the gut in experimental models. It also damages the gut mucosa. All toxigenic strains produce haemolysis when grown on fresh blood agar. This is known as kanawaga haemolysis and hence all enterotoxigenic strains are known as kanawaga-positive *Vibrio parahaemolyticus*.

The first British cases of *Vibrio parahaemolyticus* food poisonings were the passengers on an inward flight from Hong Kong. The plane stopped in India to change crew and during the flight from India many passengers developed symptoms of food poisoning. Fortunately, the new crew were unaffected in this incident. The cause was found to be *Vibrio parahaemolyticus* present in the lobster salad taken on board at Hong Kong. The crew taken on board in India had fortunately not eaten the lobster salad and so remained unaffected. The flight time from Hong Kong to the UK was longer than the incubation period: the rest we leave to your imagination. This case is of particular interest because it represents a food-poisoning incident that occurs in the food-poisoning statistics for the UK but the origin of which was half way across the world.

5.6.8   What is the role of fungal toxins in food poisoning?

Should edible fungi be referred to as mushrooms and poisonous specimens be known as toadstools? Some authors do make this distinction on the basis that the terminology is widely used. There are problems with these descriptions, however. When classifying fungi, it is not unusual to find closely related species one of which is edible while its near relative is poisonous. A good example would be the boletus mushrooms. The king boletus (*Boletus edulis*) is considered a very tasty edible fungus while its close relative the Satan's mushroom (*Boletus satana*) can cause severe indigestion when eaten. On scientific grounds, it makes little sense to differentiate these species into mushrooms and toadstools. We have, therefore, adopted the convention of referring to all the macroscopic fungi as mushrooms.

Eating mushrooms that have been freshly harvested can be a rewarding experience. What could be better than a free meal harvested with relatively little effort? There is, however, potentially a very serious problem with this apparently innocent occupation. Unless accompanied by an expert it can be all too easy to harvest a poisonous mushroom mistaking it for its edible relative. Local knowledge can be vitally important. There was a recent incident when some Dutch botanists went on a fungus foray on a visit to southern England. After enjoying a meal cooked from their harvest, several developed mushroom poisoning. Although able to identify their local dangerous fungi, they mis-identified the English poisonous mushrooms.

The folk names of dangerous mushrooms often convey their sinister character: death cap (*Amanita phalloides*), panther cap (*Amanita pantherina*), fly agaric (*Amanita muscaria*), and jack-o-lantern (*Omphalotus olearius*, also classified as *Clitocybe illudens*). None sound very appetising. There are thought to be more than 200 species of poisonous mushroom, and eating the most dangerous carries a high risk of death. Some mushrooms that have traditionally been considered suspect are now known to be edible. Their reputation probably derives from the problems caused when older specimens are consumed. When eaten fresh, these fungi may cause no problems

Almost everyone who eats large quantities of *Amanita phalloides* will die as a consequence. It is for this reason that it has earned the name of death cap mushroom. This fungus is large with a flat yellow or orange cap and white gills. Young specimens start dirty green in colour. The base of the stalk is bulbous, but this can only be seen when the fungus is dug from the ground. It is this feature that lends the species its Latin name. It and other amanita fungi produce a range of powerful mycotoxins. The first symptoms of poisoning occur six to eight hours after eating the mushroom and include nausea and vomiting and a violent diarrhoea associated with severe abdominal cramps. The onset of symptoms may be delayed depending on how much material has been eaten. As the poisoning develops, the liver becomes severely damaged and the victim consequently develops jaundice. Victims may experience breathing difficulties and the skin takes on a bluish appearance. This is described as **cyanosis** and is particularly marked around the lips. The victim then lapses into coma and, in most cases, dies. The care of victims in intensive therapy units has improved the survival of patients who have only eaten small quantities of this fungus.

Amanita verna, the gill mushroom, is probably only a variant of the death cap mushroom. It has a white cap. The panther cap (*Amanita pantherina*) is the colour of dead leaves. It is covered in scales that can be lost after exposure to heavy rainfall. It is closely related to *Amanita brunnescens* found widely in North America. Fly agaric (*Amanita muscaria*), with its bright red cap dotted with white spots, is the fungus associated with pixies and gnomes. Eating any of these mushrooms can be fatal.

The principal toxin of the death cap mushroom is amatoxin. It acts on the central nervous system, blood vessels, liver, kidneys and muscles. The amanitas produce a range of other toxins as well. Muscarine is an **alkaloid** that causes lowering of blood pressure. Its effects can be overcome by administration of atropine: a pharmaceutical product of deadly nightshade (*Atropa belladonna*). Muscarine is produced by the fly agaric (*Amanita muscaria*), from which it derives its name: other poisonous fungi produce it in greater quantities. Monomethylhydrazine is also produced by the amanita fungi and by false morels of the genus *Gyromitra*. This acts as a monoamine oxidase inhibitor in the central nervous system and is associated with a euphoric state. Derivatives of this drug have been used medically as antidepressants and one derivative, isoniazid, is used in the treatment of tuberculosis.

Ergot is another fungal toxin that acts on the central nervous system to produce hallucinations. The fungus *Claviceps purpurea* produces this toxin. This fungus infects the ovaries of rye (*Secale cereale*) and other cereal crops. It causes the ovaries to enlarge and to deform. As the rye ripens the affected seed withers. The infection develops best under humid conditions.

Besides its central nervous system effects, ergot poisoning causes severe muscle cramps, a livid skin rash, convulsions and, because of its effects on the circulatory system, **gangrene** in the limbs. Pregnant women often abort and nursing mothers can no longer feed their infants as their supply of breast milk ceases. Ergot intoxication has been referred to as St Anthony's fire since monks of the Order of St Anthony took special care of the victims of this disease. Their success may be attributed to a change in diet, excluding contaminated rye bread. Other authors have considered St Anthony's fire to be cellulitis caused by streptococci. The symptoms of the two diseases are superficially similar and it can be very difficult to interpret historical accounts of disease accurately. It has been suggested that the behaviour that precipitated the Salem witch trials in 1692 was an example of epidemic ergot poisoning. Over 300 people were tried and 20 people were executed at Salem, the last of the great witch-hunts.

As with other deadly poisons, ergot has found pharmaceutical uses. Its derivatives have been used to control haemorrhage and to induce labour to assist in childbirth. They are also used to control migraine headaches. An infamous relative of ergot is the hallucinogenic drug lysergic acid diethy-lamide: LSD. Perhaps this explains some of the bizarre creatures that popu-late the pages of mediaeval and Celtic manuscripts. These were often written in monasteries in damp areas where rye bread would have been an important component of the diet for the monks.

Mycotoxicosis is the term applied to illness caused by the consumption of fungal secondary metabolites. One of the most important is aflatoxin poi-soning. Aflatoxin was first described in 1960 when it caused an epidemic of acute enteritis and hepatitis in turkeys. Intensive investigation showed that the illness suffered by these birds results from contamination of peanut extract in their feeds. The fungus that was incriminated in this incident was *Aspergillus flavus*. The toxin was first isolated as a secondary metabolite of the fungus from which it gets its name. This toxin was subsequently found to affect humans who consume mouldy foods. The foods that are at most risk of aflatoxin production are stored under warm, humid conditions typically in tropical climates. Other species of the genus *Aspergillus* besides *Aspergillus flavus* and certain fungi of the genus *Penicillium* may also produce aflatoxin. It is typically produced when peanuts are infected with producer moulds. The extent of aflatoxin poisoning in humans is unknown but it has certainly caused a number of deaths and is associated with liver damage in a large number of people, particularly in Africa and the Indian sub-continent. In one incident, more than 400 people fell victims to aflatoxin poisoning. Aflatoxin is a powerful carcinogen associated with carcinoma of the liver. The World Health Organization now recommend the maximum level of aflatoxin in foods for human consumption. Animals are also severely affected by aflatoxin as the first incident illustrates. People who feed wild birds on mouldy peanuts may do much more harm than good. If you would not fancy eating the mouldy peanuts you put out in the garden, why inflict them on the poor birds? In addition to aflatoxin, there are now a number of other mycotoxins that are produced as secondary metabolites of fungi growing on food. These may have various clinical effects including nerve damage and stunting of growth.

5.6.9 What viruses cause food-borne illness?

Although viruses are a very common cause of gastroenteritis, especially in very young children, virus-associated food poisoning is not common. The food-borne virus illnesses are most often derived from eating shellfish that have been harvested from waters contaminated with raw sewage. The most common cause of virus-associated food poisoning is the small fastidious viruses that are associated with eating filter-feeding bivalve molluscs such as cockles and mussels. Hepatitis A virus can also be acquired from eating shellfish. A particular problem is that some people regard raw oysters as a delicacy and as an aphrodisiac. These shellfish suck their microscopic food from the sea by pushing large quantities of water over their filter gills. An oyster of modest size can pump many times its own volume of water per hour. Small particles of food become trapped in the mucus lining of these gills and the beating of cilia moves this mucus to the opening of the gut proper. This method of feeding lends itself to the concentration of virus particles and these are subsequently eaten together with the shellfish. It is in this way that they can propagate human gastrointestinal infection. In one incident, 61 people in Florida were infected with hepatitis A virus after eating oysters from a single affected oyster bed.

5.6.10 What are the pre-disposing factors in food poisoning incidents?

There are many factors that pre-dispose food to bacterial contamination. Consequently many factors may contribute to food-poisoning incidents. Different bacteria are more likely to cause problems in certain situations than others, depending upon the biology of the organisms. More than one factor may contribute to a food-poisoning incident.

The most common factor contributing to food poisoning is preparing food too long before its consumption. Nearly one in four food poisonings have this as a contributory factor. It is particularly involved in incidents associated with *Bacillus cereus* and *Clostridium perfringens*, where the spores survive cooking and then germinate so that bacteria can multiply in the cooked food. The second highest risk factor is storage of food at ambient temperature. Again, this permits multiplication of food-poisoning organisms, enabling an infective dose to be achieved. Inadequate cooking fails to destroy all bacteria within the food, particularly the Gram-positive sporing bacillus *Clostridium perfringens*. Inadequate reheating of food causes similar problems to inadequate cooking and is associated with a particular risk of *Clostridium perfringens* and *Bacillus cereus* food poisoning. Contaminated processed foods are more likely to be associated with salmonellas and *Staphylococcus aureus* rather than sporing bacteria. Salmonellas are linked with food-poisoning incidents where previously frozen food has not been adequately thawed prior to cooking, or in incidents where foods become cross-contaminated. In cross-contamination, infected food comes into contact with previously uninfected food, inoculating bacteria onto the new food. This is particularly serious if the cross-contaminated food is eaten without further cooking. Raw meats should always be kept away from cooked products etc. in a refrigerator and should be placed on a lower shelf than other foods to prevent drips from contaminating the contents of the refrigerator. Infected food handlers are a common source of *Staphylococcus aureus* food poisoning; they rarely cause salmonella infections.

6

The human commensal flora

6.1 What constitutes the resident and transient flora of humans?

It has been estimated that the human body contains 10^{14} cells. Of these, 90% are not of human origin. They represent the microbes of our **commensal flora**. The term commensal is partly derived from the Latin word *mensa*, meaning table. Commensal organisms are considered to share their food from a common table; one that we provide as human hosts. Different anatomical sites are associated with a flora peculiar to each location.

During our time in the womb we live in a sterile environment, protected on one side by the placenta and on the other by the amniotic sac. From the moment of birth, however, we are subjected to a huge array of microbes. The first organisms that we as babies come across are those present in the birth canal. During birth we inhale, swallow and acquire on our surface a vast diversity of microbes as a result of contact with the new environment. This process will continue throughout our lifetime. If the organisms with which we come into contact find themselves in a suitable ecological niche, whether on an internal or an external surface, they will multiply and form complex communities. They will interact with each other and with their human host. This process requires the microbes to adhere to the host as an initial step in the colonisation process and then the microbes must multiply. Those microbes that are able to achieve this and that are particularly suited to their new location may form long-term, stable, interdependent relationships with their neighbours and with the human that harbours them. Such organisms constitute our **resident commensal flora**.

Not all microbes that come into contact with humans find us a hospitable habitat. Some fail to establish a foothold at all. Others do manage to colonise humans but they do not become established in their new home, either because of competition from the resident flora, or because of their susceptibility to the various host defences. These organisms do not make a permanent home on us and are referred to as the **transient flora** to reflect the temporary nature of the relationship with their human host.

A third group of microbes form a relationship with humans. After establishing themselves on or in the host, initiating an infection, they go on to do damage. In its most severe form the injury caused may result in the death of the human. In the vast majority of cases, host defences are mobilised to limit the harm and, generally, to eliminate the offending microorganism. Such microbes are known as **pathogens.**

Commensals are considered to coexist in harmony with their host, whereas pathogens are perceived to do harm. The distinction between the two groups is, however, not absolute. *Staphylococcus saprophyticus* is frequently found as a constituent of the commensal flora of the skin. During energetic sexual intercourse this bacterium may gain access to the female urinary tract. There it may flourish, not having any competing flora to limit its growth. In this way, *Staphylococcus saprophyticus* is one of the most common causes of 'honeymoon cystitis'. Conversely, pathogens may be isolated from healthy individuals who have never exhibited symptoms of infection. Such people are referred to as carriers, and they may act as important sources of infection for others. A man employed to lay water pipes in Croydon in 1937 was a typhoid carrier who shed *Salmonella typhi* in his urine. He was in the habit of urinating on the work-site and, as a result, he contaminated the town's water supplies. This led to 310 cases of typhoid and 43 deaths from the disease. The carrier remained perfectly healthy and had never exhibited symptoms of typhoid.

Microbial access to deep tissues is only transient, and in healthy individuals any organisms that do enter such sites are rapidly removed by the various host defence mechanisms. Normally, **cerebrospinal fluid** is sterile and no microbes should be found in muscles, joints, bones or connective tissues. Our blood, kidneys, liver, urinary tract and spleen all have efficient mechanisms for the removal of microorganisms. Established microbial populations may, however, be found on the skin, nasopharynx, upper respiratory tract, mouth and lower gut, as well as in the vagina. Throughout life, the constituents of the normal flora vary in type and number, and different individuals may display marked differences in their resident microbial populations. Despite this, generalisations concerning the human commensal flora are possible.

6.2 What constitutes the commensal flora of the human skin?

The skin is a highly complex organ that provides a variety of ecological niches for microbial colonisation. It is also the first line of defence against infection and as such possesses a variety of mechanisms that have evolved to minimise microbial overgrowth. These include the production of sebum, a fatty substance inhibitory to many microbes. The skin also secretes lysozyme, which attacks and weakens bacterial cell walls. Additionally the pH of the skin is lower than can be tolerated by many microbes. Furthermore, the resident commensal flora of the skin contributes to its defences against further microbial colonisation. Propionibacteria can metabolise the components of sebum to produce unsaturated fatty acids that have a marked antimicrobial activity. Skin is, therefore, an inhospitable environment. Despite these defences, coagulase-negative staphylococci, micrococci, corynebacteria, propionibacteria, lactobacilli and yeasts may all be found in the resident flora of the skin, particularly in sweat glands and hair follicles. Bacteria occur in microcolonies on the skin but these are not evenly distributed over the body. The head, armpits, groin, hands and feet are the most heavily populated sites, where microbial counts of many thousands per square centimetre may be found. Males are more heavily colonised than females.

As well as bacteria, fungi and even mites form part of the resident commensal flora of the human skin. The fungus *Malassezia furfur* was formerly known as *Pityrosporum ovale* or *Pityrosporum orbiculare* depending upon its morphological type. It and similar fungi may be isolated from the skin of many individuals, as can yeasts of the genus *Candida. Candida albicans* does not normally colonise the skin. All humans can act as hosts to skin mites such as *Demodex folliculorum* and *Demodex brevis*. These animals survive on a diet of dead epithelial cells supplemented with sebum.

The eye is covered with a specialised skin bathed in tears. These wash the surface of the eyes continuously. Few microbes can survive in these conditions. Nevertheless, corynebacteria, especially *Corynebacterium xerosis*, can establish themselves as resident commensals. Other microbes may be present on the surface of the eye in very small numbers but their presence is transient.

Microbial interactions with skin secretions, particularly in the armpits and groin, lead to the formation of volatile fatty acids responsible for body odour. To combat this, deodorants have been formulated containing substances with a selective activity against Gram-positive bacteria such as are found on the skin. Use of such products reduces the amount of aromatic volatile products at these sites. It can also pre-dispose the consumer to colonisation and possible infection with Gram-negative bacteria, which may flourish without competition from the resident Gram-positive flora.

There was considerable excitement when the brevibacteria found in the particularly smelly Limburger cheese were also found on the feet. This was thought to be the reason that people may be afflicted with smelly feet. The story is not that simple, however. Careful examination shows little correlation between the degree of smell of the feet and the number of brevibacteria that they carry. The habit of enclosing our feet in shoes does contribute to this problem and people who wear heavy footwear for long periods tend to be more prone to cheesy feet than others. Such people are also prone to develop pitting of the toughened skin because of the damp environment that is generated inside shoes. This pitting is associated with the activity of another foot commensal: *Micrococcus sedentarius*. This bacterium, like the brevibacteria, can produce large quantities of methanethiol, a volatile chemical with a cheesy smell. Again, numbers of *Micrococcus sedentarius* do not correlate with the degree to which feet smell. The best correlation between cheesy feet and bacteria is with the aerobic coryneforms and the staphylococci. The higher the numbers of these bacteria on the feet the more likely are the feet to be smelly.

The commensal flora resident on the skin may become dispersed into the environment as epithelial cells become sloughed off the surface of the skin. This is of particular importance because 10–40% of healthy adults in the general population and up to 90% of hospital staff carry *Staphylococcus aureus* on the skin around the nose or groin. *Staphylococcus aureus* is one of the most common causes of surgical wound infection and almost all healthy babies born in hospital become colonised with *Staphylococcus aureus* within a week.

6.3   What constitutes the commensal flora of the human alimentary tract?

The mouth is the route by which we take in food. It also provides easy access to a wide array of microbes, some of which may become part of our transient microbial flora. Saliva provides a washing mechanism that will help to prevent microbes from becoming established within the oral cavity. We swallow an average of 30 times per hour. Besides providing mechanical protection, saliva also contains digestive enzymes and a number of antimicrobial agents including secretory immunoglobulin A (IgA), lysozyme and lactoferrin. We also indulge in oral hygiene: brushing our teeth and dental flossing, for example. With all this protection it could be thought that the mouth would not have an extensive flora. This is untrue. The oral cavity provides a number of distinct anatomical sites and these are colonised by a huge array of microbes.

The various surfaces of the oropharynx support a very mixed collection of microbial populations. Even structures that lie in contact with each other, like the tongue and the teeth, are colonised with different types of microbe. *Streptococcus salivarius* is found on the soft epithelial surfaces of the mouth, whereas the hard surfaces of the teeth are colonised by *Streptococcus mutans* and *Streptococcus sanguis*. The gingival crevices between the gums and teeth harbour miscellaneous bacteria including species of the genus *Bacteroides*, *Fusobacterium* species and spirochaetes, forming intricate parasitic interactions. Saliva contains mostly α-haemolytic streptococci, most notably *Streptococcus salivarius*, although a variety of other microbes may also be found.

The stomach is not normally heavily colonised, though a few acid-tolerant lactobacilli may be found there. This is because the acidic nature of the gastric secretions is actively bactericidal. *Helicobacter pylori*, linked with ulcer formation, can exist in these conditions because it produces a highly active urease. Also, the waxy cell wall of mycobacteria enables them to survive the acid conditions of the stomach. Microorganisms entrapped within food may be recovered from the stomach after a meal. These contribute to the transient flora of the stomach.

The upper portion of the small intestine, comprising the duodenum, jejunum and upper ileum have no resident flora. The digestive secretions, bile acids, intestinal mucus and secretory antibodies make this too hostile an environment to permit colonisation. Active peristalsis probably also contributes to the absence of a resident flora. Evidence for this is found in the observation that bacterial overgrowth occurs if peristalsis is prevented, for example by the formation of an intestinal 'blind loop'. The microorganisms that are found in the overgrowth are similar to those found in the large intestine.

At the lower end of the ileum, an abundant flora may be found. In the upper reaches, this flora may contain microbes that are also found in the mouth, but as the large intestine is approached, there is a gradual transition to a faecal type flora. Bacteria are very numerous in the large intestine and obligate anaerobes such as *Bacteroides* species are far more numerous than facultative organisms like *Escherichia coli*. Obligate anaerobes outnumber coliforms by at least 100:1. This ratio may be a considerable underestimate, given the difficulty involved in cultivating some obligate anaerobes. It has been suggested that the metabolic activity of the bowel microflora is equivalent to that of the liver. Intestinal bacteria are responsible for the breakdown of bile acids. They also provide vitamin K, a vitamin that has to be supplied in the diet of germ-free animals.

The effects of the bowel flora are not always benign. There have been suggestions that metabolism of the bowel flora generates carcinogens that are implicated in the development of carcinoma of the bowel but definitive evidence has proved elusive. There is also the problem of flatulence. The problems of lactose intolerance have already been discussed for those people who lack the lactase gene and cannot drink unfermented milk as a result. Similarly, foods such as baked beans contain oligosacccharides that we humans cannot digest but that our commensals can. Eating too many baked beans leads to antisocial products that, poetically, are reputed to be good for your heart.

Along with causing the problem of flatulence, the bowel flora makes its own contribution to controlling it. It has been calculated that gas released from the activity of bowel microbes ought to reach about 24 litres per day. Common sense shows that this does not happen and experiments have shown the volume of flatus released per day by an average person is only about 1 litre. The remaining gas is thought to be eliminated as microbes in the bowel utilise the volatile products of their neighbours.

6.4   What constitutes the commensal flora of the human upper respiratory tract?

The respiratory tract is an anatomically complex structure constantly exposed to microorganisms from the air we breathe. Microbes inhaled in the air often become entrapped on hairs inside the nostril, or upon the mucous membranes of the turbinate baffles found further up the nose. The microflora of our nostrils resembles that of the skin and, like the skin, the upper respiratory tract is colonised by staphylococci, micrococci and corynebacteria. As well as these organisms, the warmer, moist environment of the upper respiratory tract provides a haven for a large, complex and varied flora of Gram-negative as well as Gram-positive bacteria. Among the bacteria found in the upper respiratory tract are *Streptococcus, Moraxella, Neisseria* and *Haemophilus* species. *Streptococcus pyogenes*, which may cause tonsillitis, may be found as part of the commensal flora of the nose in healthy carriers.

The paranasal sinuses are not normally colonised, neither are the larynx and lower respiratory tract. The lungs are kept free from colonisation by the action of the 'mucociliary escalator'. Mucus traps any particles, including microbes, that may intrude upon the lower respiratory tract. Cilia in the lungs beat to drive a stream of mucus from the alveoli through the bronchial tree and up the trachea. As a result, mucus accumulates at the top of the throat and periodically it is swallowed. The extremely acid pH of the stomach renders it inhospitable to most microorganisms that arrive there as a result of the activity of the mucociliary escalator.

6.5 What constitutes the commensal flora of the human genital tract?

The urethra has an effective cleansing mechanism. Regular flushing with urine combined with the activity of secretory antibodies and, in males, prostatic secretions all help to eliminate colonisation. For the majority of its length the urethra is sterile, but a sparse flora, comprising staphylococci, Gram-negative cocci, corynebacteria and mycoplasmas, is located in the distal portion. The external genitalia in many people are colonised with *Mycobacterium smegmatis*. This bacterium may contaminate urine samples and confuse the laboratory diagnosis of renal tuberculosis.

In pre-pubescent and post-menopausal women, the vaginal flora comprises enterococci, coagulase-negative staphylococci, coliform bacilli and corynebacteria. During the reproductive years, the vaginal flora is dominated by lactobacilli, sometimes referred to as Döderlein bacilli, in a mixed flora also containing yeasts, corynebacteria and mycoplasmas.

6.6 What is the role of the human commensal flora?

Louis Pasteur believed that it was not possible for animals to exist without a commensal flora and early experiments to raise germ-free, or gnotobiotic, animals met with failure. The term 'gnotobiotic' is taken from two Greek words '*gnosis*' meaning knowledge and '*bios*' life. Gnotobiotic animals are often used in studies of the commensal flora in which known collections of bacterial species are introduced to determine the effect of particular bacteria on the host: hence the term gnotobiotic.

We now know how to rear gnotobiotic animals. Although colonies of such creatures can be maintained by natural reproduction, the animals require a diet that is supplemented with very high levels of vitamin B and vitamin K. Vitamin K is a vitamin not usually required by healthy animals since it is a metabolic product of gut bacteria such as *Bacteroides* species and *Escherichia coli*. Gnotobiotic coprophagous (dung-eating) species such as rabbits have a greatly enlarged caecum, the part of the gut where digestion of plant cell wall material takes place. All germ-free animals have a poorly developed immune system. This points to a role of the normal flora in fulfilling the nutritional demands of the host and in stimulating the immune system. When gnotobiotic animals are reared on a properly supplemented diet, however, they grow faster and live longer than conventional animals carrying a resident microflora.

The commensal flora may play a significant role in denying pathogens access to their target organs. This may be simply a result of denying the invader access to its target site for adhesion. Alternatively, the constituents of the normal flora may actively produce substances that inhibit the growth of other microbes or even kill them. An important example of microbial exclusion is afforded by antibiotic-associated pseudomembranous colitis. Following administration of broad-spectrum antibiotics, the anaerobic flora of the colon becomes depleted, leaving a biological vacuum. This permits the potential for colonisation by a toxigenic strain of *Clostridium difficile*, which may go on to cause a serious, life-threatening diarrhoeal disease. The infectious dose of salmonella may also be influenced by the presence of the commensal flora. It usually takes about a million organisms to initiate a clinical salmonella infection. In volunteers fed salmonella together with oral streptomycin, an antibiotic that killed potential competitors in the commensal flora, the infectious dose was dramatically reduced. The gut is not the only site where such interactions have been observed. The flora of the female genital tract provides another example. The presence of high numbers of lactobacilli in the vaginal flora is thought to reduce the incidence of gonorrhoea, since women who develop the disease are less likely to have lactobacilli in their normal flora. Furthermore, recovery of lactobacilli is highest during the first part of the menstrual cycle and it is during this phase that recovery of *Neisseria gonorrhoeae* is at its lowest.

The commensal flora is an important source of infection for the human host. Infections caused by microbes derived from our commensal flora are referred to as endogenous infections. These may range from trivial conditions such as boils through to life-threatening infections. An example is bacterial **endocarditis**, where α-haemolytic streptococci from the mouth or coagulase-negative staphylococci from the skin gain access to the bloodstream and infect the interior of the heart. Coagulase-negative staphylococci may also produce large amounts of slime and thus can stick tightly to plastics. These bacteria comprise one of the principal causes of infections associated with the use of shunts to relieve **hydrocephalus**, a condition in which the pressure of cerebrospinal fluid within the brain is too high. Urinary catheterisation also carries a higher than normal risk of Gram-negative septicaemia, which can be fatal. The bacteria responsible are most often found as part of the faecal flora. *Clostridium perfringens* is often found on the skin of the legs. This bacterium is the causative agent of gas gangrene. This condition is often associated with the lower limbs.

Not all endogenous infections are life-threatening, at least in the short term. The resident commensal flora of the mouth is responsible for the formation of dental plaque. *Streptococcus mutans, Streptococcus salivarius* and *Streptococcus mitior* elaborate polyglucans from the sucrose present in our diet. *Streptococcus salivarius* also produces polyfructans. These extracellular bacterial polysaccharides interact with salivary glycoproteins to form a glutinous film in which bacteria may become enmeshed. Plaque formed in this way adheres firmly to tooth surfaces. When it first forms, the major bacterial components of plaque are streptococci, but if it is left undisturbed, filamentous bacteria and obligate anaerobes rapidly outnumber the streptococci. The acidic metabolic products of these organisms are extruded onto the enamel of the tooth whence they initiate dental **caries** – tooth decay.

6.7   What factors affect the human commensal flora?

Microbes of the commensal flora are in a constantly changing dynamic equilibrium with their human host. Strains are continually being displaced and replaced by other strains. In this way, the commensal flora adapts to changes that occur in the host. This process may be exemplified by changes occurring in the vaginal flora throughout life and by comparing the bowel flora of breast-fed and bottle-fed babies.

At birth, the vagina is sterile but it rapidly becomes colonised with microbes found on the skin, together with enterococci. Within days, lactobacilli predominate as glycogen is secreted by the vaginal epithelium. These factors combine to cause an acidic pH. All this is thought to be in response to maternal hormones, which are still circulating in the infant girls. The presence of the lactobacilli is only temporary and their numbers wane as the influence of the maternal hormones declines. A mixed flora of enterococci, coliforms and corynebacteria displace the lactobacilli. The environment within the vagina then becomes alkaline. With the onset of puberty, glycogen is again secreted, the pH of the vagina falls in response to hormonal changes and lactobacilli are then predominant in the mixed vaginal flora. At the menopause, the flora reverts to that of the pre-pubescent vagina.

A variety of factors influence the composition of the faecal flora. Maternal antibodies present in breast milk, together with the iron-binding protein lactoferrin, exert an inhibitory effect on coliform bacteria such as *Escherichia coli*. In consequence, babies fed on breast milk produce faeces with a low pH and a flora that predominantly breaks down sugars, which includes bifidobacteria and enterococci. Very few clostridia, bacteroides and coliform bacteria are found in the faeces of babies fed on breast milk. Processed bottled milk does not contain such effective inhibitors. Consequently, bacteroides, clostridia and coliforms dominate the faecal flora of bottle-fed babies. As a result, the physical appearance and smell, as well as the microbiology of the faeces, of bottle- and breast-fed babies are significantly different. The low pH and oxygen potential of the faeces of breast-fed babies also helps to contribute to the overgrowth of bifidobacteria. When breast-fed babies are introduced to mixed feeds, bifidobacteria subside and a putrefactive flora, similar to that of bottle-fed babies and adults, soon appears.

6.8 Do viruses form part of the human commensal flora?

Healthy humans may harbour virus material that apparently does no harm. This material may, perhaps, be considered to form part of the human commensal flora. During carriage, healthy individuals are not making infectious virus particles. Under certain circumstances, the virus may, however, become reactivated, causing a clinical illness. The viruses also return to an infectious state.

Members of the herpesvirus family provide the prime example of viruses that may become latent. Upon primary infection, the varicella-zoster virus causes chicken pox, characterised by a generalised vesicular skin rash. The virus enters the central nervous system and there it enters a latent phase. In this state the virus DNA can persist for many years. If an individual then suffers, for example, a diminished **cell-mediated immunity** for any reason, the virus may reactivate and pass down the peripheral nerves to cause shingles. Clinically, shingles presents as a vesicular eruption of the skin attached to the peripheral nerves down which the reactivated virus has travelled. Virus particles from shingles lesions may then infect susceptible individuals, causing chicken pox. Cold sores and genital herpes are caused by a similar reactivation of the latent form of herpes simplex viruses types 1 and 2. Reactivation can occur in the absence of a clinical response. If, however, lesions result from the reactivation then this is referred to as a recrudescence. Cytomegalovirus and Epstein–Barr virus, both herpesviruses, and adenoviruses can persist in lymphoid tissue. The papovaviruses BK and JC also persist in many healthy individuals worldwide. BK virus may reactivate following immunosuppression for renal transplant surgery, causing problems with the transplanted kidney and its associated urinary tract. JC virus is associated with a neurological degenerative disease, PML or **progressive multifocal leukoencephalopathy**, that occurs in people who are severely immunocompromised, particularly patients suffering from lymphoma or the acquired immunodeficiency syndrome (AIDS).

Viruses can exist in two forms: latent and persistent. As such they may be considered to form part of the commensal flora of humans but they do differ from higher commensals. Many bacteria and higher organisms in the commensal flora are capable of causing an infection if they gain access to places where they are not usually present. They do not alter in their structure when causing **opportunistic infections**. Viruses may exist in a truly latent form where infected cells do not produce infectious virus particles unless the virus is reactivated. The herpes family provide examples of truly latent infections. Alternatively, they may exist as a persistent infection with infectious particles being continually produced from infected cells. In the latter case there may or may not be a clinically apparent response. Examples of persistent viruses include hepatitis B, hepatitis C and the human immunodeficiency virus. Individuals with persistent infection are usually infectious and are referred to as carriers.

7

Microbial infections

7.1 How do microbes cause disease and how do we defend ourselves from infection?

The vast majority of microbes that humans encounter on a daily basis do no harm. Indeed, the microbes that constitute our commensal flora are, on balance, beneficial under most circumstances. Only a minority of microbes can interact with humans to cause disease. These are known as pathogens from the Greek word *pathos*, meaning 'suffering'. Some pathogens cause mild illnesses: others are responsible for life-threatening infections. Some infections are chronic, developing slowly over many months or years: others may be rapidly fatal. **Virulence** is the term used to describe the degree to which an organism can cause disease. This term is derived from the Latin word *virus*, meaning a poison. Some authors do use the terms 'pathogenicity' and 'virulence' interchangeably. Other authors prefer to reserve the term virulence for a quantitative description of the degree to which a microbe can damage its host.

Infections spread from a source known as the **reservoir of infection**. These may be human beings, as with *Salmonella typhi*, animals, as with *Salmonella typhimurium*, or the environment. Soil is the reservoir of infection for tetanus, caused by *Clostridium tetani*, and water for Legionnaire's disease, caused by *Legionella pneumophila*. The **source of infection** is the individual or location from which an infection is acquired. If the source of infection is the patient's commensal flora, the infection is said to be **endogenous**, but if it is acquired from elsewhere the infection is said to be **exogenous**. Vectors of infection may spread infections. Fleas that live on rats are responsible for the spread of plague. If the vehicles of infection are inanimate, they are known as **fomites**. When apparently unrelated cases of infection occur within a population, they are said to be **sporadic**. If a disease is continuously present in a community it is **endemic**. **Epidemics** occur when the number of cases of a particular infection rises significantly above the endemic level. **Pandemics** are epidemics that cover the whole world.

Infections may be transmitted in a number of ways. These include from person-to-person, air-borne, water-borne, food-borne and insect-borne infection. *Legionella pneumophila,* for example, may be spread in air-borne water droplets shed from contaminated shower heads. Person-to-person spread can occur through several routes. Many respiratory and systemic infections are spread by the inhalation of infectious droplets. Diarrhoea provides a liquid medium in which gastrointestinal pathogens may easily spread by the faecal–oral route. Some of the most vulnerable of human pathogens are those causing venereal diseases. These have evolved to spread during sexual intercourse at moments where humans share a high degree of intimacy. This allows the pathogens to pass between individuals without being exposed to a harsh external environment. Occasionally, person-to-person infections may spread by direct inoculation. This can occur when, for example, a person is bitten. Alternatively, direct inoculation can result when intravenous drug users share contaminated needles with one another or in individuals who may be exposed to contaminated needles, for example in tattoo parlours, acupuncture clinics and some hospitals. Most infections that spread from one individual to another are said to spread horizontally, through a population. When, however, a mother infects her baby the infection is said to be spread vertically. Such transmission is from one generation to the next. Vertical transmission of infection can occur during the development of the fetus, at birth or through infected breastmilk. Water- and food-borne infections have previously been described in Chapters 3 and 5, respectively. Historically, the most important insect-borne infection is the plague. Today, malaria, dengue and yellow fever provide other important examples.

Some pathogens can cause disease by intoxication. In such cases, live microbes do not need to enter the body. Rather, the disease state is a response to the action of a toxin. These may be relatively mild and self-limiting, as with staphylococcal food poisoning, or more serious: in the case of botulism, they are generally fatal. Other pathogens colonise the body surfaces. Examples include the viruses that cause the common cold and also *Vibrio cholerae*, the cause of cholera. Very few people die from colds but there is a significant mortality associated with cholera. Yet other microbes penetrate into the submucosal layers. These include the shigellas: causes of bacilliary dysentery.

Finally, there are microbes that penetrate the body. These pathogens are exposed to our full range of defences and must circumvent all our antimicrobial mechanisms to complete their lifecycles. Some, such as *Salmonella typhi*, have even evolved to live inside white blood cells: the very cells that we use to protect ourselves from infection. All viruses, being obligate intracellular parasites, must replicate intracellularly. Some virus infections can be clinically inapparent: others are rapidly fatal. These examples illustrate how difficult it can be to make predictions regarding the severity of disease, even when we know how far microbes penetrate their hosts.

Animal models have been used in the past in an attempt to measure the virulence of an organism. Although some animal models mimic human infection well, as with ferrets and influenza, these studies have not always been satisfactory. For example, *Salmonella typhi* is a pathogen that is associated with serious human infection. It causes an enteric fever with the bacteria living inside circulating leukocytes. In a mouse model, however, the bacterium colonises the gut only, where it causes a mild gastrointestinal disease. In contrast, *Salmonella typhimurium* causes human gastroenteritis. In the mouse, however, it causes enteric fever. The Latin name *typhimurium* means 'mouse typhoid'. The genus *Salmonella* illustrates another feature of pathogenicity. *Salmonella typhi* is a human pathogen. It is not found in other species either causing disease or as a commensal bacterium. Other salmonellas are, however, widely distributed. About 3% of healthy human adults carry a salmonella as part of their gut flora and salmonellas can be isolated from a wide range of mammals, birds and reptiles. In some they cause disease; in others they act as commensals.

Even the degree to which individuals respond to infection may be highly variable. This is illustrated by the Lubeck disaster. During 1926, children in Lubeck in Germany were accidentally vaccinated with a virulent strain of *Mycobacterium tuberculosis* rather than the attenuated vaccine strain. Of the 249 children given the virulent vaccine, 76 died. The remainder developed only minor lesions. Each child was vaccinated with a standard inoculum. Consequently, the different responses of the children to the vaccine reflect differences in the host response to infection. Another problem with the study of infection is that in some cases a single pathogen is associated with a single disease process: *Yersinia pestis* causes plague. Other microbes cause a range of diseases. Diseases caused by *Staphylococcus aureus* range from trivial conditions like staphylococcal food poisoning, impetigo and boils, to much more serious illnesses such as osteomyelitis, endocarditis and toxic shock syndrome. These can be life-threatening. The study of pathogenesis is fraught with problems. Very early in the development of microbiology as a science, Robert Koch attempted to define the criteria needed to illustrate that a particular microbe is, indeed, the cause of a particular infection. These have been refined over the years and are collectively known as **Koch's Postulates**.

1. The agent must be present in every case of the disease;
2. the agent must be isolated from the host and grown in artificial culture;
3. the disease must be reproduced when a pure culture of the agent is inoculated into a healthy susceptible host;
4. the same agent must be recovered again from the experimentally infected host.

In practice, even classical infectious diseases may not fulfil Koch's Postulates. The bacterium that causes syphilis, *Treponema pallidum*, cannot be grown in artificial culture. In other cases, for example lethal infections with humans as the only susceptible host, establishing Koch's Postulates would be unethical.

For any microbe to survive it must find a suitable ecological niche, colonise and exploit its habitat, multiply in numbers and disperse to a new and suitable habitat. With pathogens, humans provide a variety of different ecological niches. Each anatomical site may be exploited by its particular pathogens. The symptoms of infectious diseases are a consequence of the multiplication and spread pathogens undergo to complete their lifecycle. In order to colonise the host successfully, pathogens may elaborate a variety of **virulence factors**. Important examples include toxins that damage tissues directly, adhesins that aid colonisation of the surface of a new host and capsules that defend the microbe from our defences. All the bacteria that cause meningitis are, for instance, protected by a capsule.

Control of infection would appear to be simply a matter of identifying the mode of spread of the pathogen and interrupting the cycle of infection replication and spread. The wit who wrote the following advice on a lavatory wall had obviously learned how to break a cycle of infection: '. . . half the girls in this college have TB, the others have VD. Sleep with the ones who cough'. In reality breaking the cycle of infection is often easier said than done. Only in the case of smallpox has an infection been eradicated entirely from the world. This was achieved through a vaccination programme that effectively denied the pathogen a reservoir of susceptible hosts in which to replicate. Protection for the individual is provided by what is referred to as herd immunity. A large proportion of the population are rendered insusceptible to infection, reducing the opportunity for the pathogen to spread from host to host. Herd immunity levels for most infections need to be in the order of at least 80% of the population if the vaccine programme is to be effective.

Our innate defences naturally protect us from infection. Human skin is an inhospitable environment. There is relatively little moisture on the skin and where the skin is wet this is because of sweat. Sweat is an excretion with a very high salt content: that is why dogs like to lick sweaty skin. The concentration of salt present provides additional problems for microbes trying to colonise skin. Furthermore, the commensal flora that manages to live on the skin can metabolise the fatty secretions in sebum to yield free fatty acids. These products can be potently antimicrobial. Skin is also a hard and multi-layered structure with flakes being sloughed off continually. Mucous membranes are not as tough as skin but they may be much more heavily colonised with an indigenous flora. These commensal microbes physically prevent pathogens from gaining access to sites that they would otherwise colonise. Alternatively, mucous membranes have elaborate cleansing mechanisms such as the mucociliary escalator that keeps the lungs sterile. Additionally, mucous membranes are protected by the activity of roving phagocytic cells as well as by secretory antibodies and other chemicals.

Once a pathogen has penetrated our surface barriers, we are still protected by an array of defences. Phagocytic leukocytes are important in providing a cellular defence against infection. These are attracted to invading cells by a little-understood chemotactic mechanism. The invading microbe is then subjected to phagocytosis. Once inside the phagocyte, the lysosomes within the cell discharge antimicrobial chemicals. These include lysozyme to break down the bacterial cell wall and lactoferrin to bind any intracellular iron and thus prevent microbial growth. Intracellular killing is effected by the generation of superoxide and peroxide radicals. These are rapidly toxic to microbial cells. There are leukocytes known as natural killer (NK) cells that attack cells infected with viruses. Infected cells carry glycoproteins on their surfaces and these act as recognition signals for NK cells to attach to the infected cell and then kill it.

Besides a cellular response to infection, we are also protected by our complement system. This is a collection of proteins that act together to produce a cascade response. Even a weak signal can be amplified in this way to elicit a strong response. The complement system has two major effects. It can act directly on invading microbes or it can act in association with antibody to cause cell lysis. It does so by puncturing holes in the microbial cell membrane. The complement system also binds to the outside of microbes, making them much easier for phagocytes to engulf. This activity is referred to as **opsonisation**, since the Greek word *opson* means cooked meat, implying something good to eat. The antimicrobial effect of complement is enhanced by a variety of other proteins. These include interferons, proteins that are important in protecting against virus infection. Additionally, a range of 'acute phase proteins' defends us. These include C-reactive protein, tumour necrosis factor and a number of interleukins. These all act to protect us from invading pathogens.

Fever is an important response to infection mediated through various cytokines. It may be unpleasant as an experience but it is a very ancient response to infection. It probably even pre-dates warm-blooded animals. If lizards are experimentally infected and held at a high temperature, they are more likely to recover than if held at a low temperature. Furthermore, lizards that are infected and that are given a range of temperatures at which they can remain are more likely to settle in high-temperature environments than uninfected controls. In at least one case, fever caused by one infection can cure another. One old cure for syphilis, used before penicillin became widely available, was to give the patient malaria. This causes regular, predictable bouts of fever. *Treponema pallidum* cannot tolerate repeated fevers.

The above defences are non-specific and are unable to differentiate between pathogens. We are, however, also protected from infection by specific defences. Once we are invaded by a pathogen, our specific defences attempt to eradicate the organism and to protect us from subsequent attack. The cells responsible for these defence mechanisms are lymphoid cells, produced in the bone marrow and found in lymph nodes, the thymus and spleen. Lymph nodes include the tonsils, adenoids and the Peyer's patches within the gut. Lymphoid cells consist of B and T lymphocytes. B lymphocytes protect us by secreting specific antibody molecules that interact with particular pathogens to destroy them. T lymphocytes interact directly with pathogens. They are referred to as the cell-mediated arm of the immune response. Other T cells are involved in helping many aspects of the specific immune response, for example T helper cells. It is these cells that are killed by the human immunodeficiency virus (HIV) thus severely compromising the immune system in AIDS patients. Most specific defence mechanisms produce memory cells during the first exposure to an infection. This is why we do not, in general, succumb repeatedly to the same infection.

Antibodies are also known as **immunoglobulins**. They are made by cells derived from the bone marrow and known as B cells. Two important classes of immunoglobulin are found circulating in plasma, where they make up about 20% of the plasma protein. These are IgM and IgG: Y-shaped glycoproteins that can interact with specific sites on antigens known as **epitopes** (Fig. 7.1). They comprise heavy and light chains. Much of the amino acid sequence is conserved and this is known as the constant region of the molecule. At either end of the Y branches, however, the amino acid sequence is

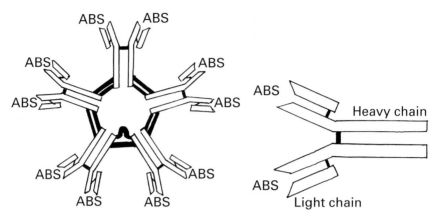

Fig. 7.1. The structure of IgM (left) and IgG (right) IgM comprises five immunoglobulin units in a pentameric molecule: IgG is a single immunoglobulin unit. Each immunoglobulin unit has two light chains and two heavy chains (the longer sections). These are joined by disulphide bridges. At the antigen-binding site, the amino acid structure is highly variable, to allow binding to different epitopes. Behind this variable region in both heavy and light chains, the amino acid sequence is constant.

variable. It is this area that binds with the epitope and this gives individual antibodies their specificity.

The first antibody to appear in response to infection is IgM. This has a pentameric structure and is the basis of our primary immune response. Because of its pentamer structure it has ten binding sites, but nonetheless it is not as effective as IgG. IgG is a monomeric molecule that is produced in large quantities as a secondary immune response. It is the antibody class that is important in maintaining our immunity over time since it interacts very tightly with its epitope. This is referred to as having a high affinity for its binding site. It is the only immunoglobulin to cross the placenta and maternal IgG is important in protecting newborn babies from infection. There are three other classes of immunoglobulin recognised. IgA is the principal antibody found in secretions such as saliva and tears and in the mucus secretions of the respiratory, intestinal and reproductive tracts. It is generally seen as a dimer structure. IgE is involved in allergic reactions; little is known about the function of IgD.

Although protection from infection is conferred by antibody, cell-mediated immunity is often of primary importance in recovering from an infection. Cell-mediated immunity is of particular importance in protecting from parasite infestations, many virus infections and in chronic diseases such as tuberculosis. Central to the cell-mediated immune response are the T lymphocytes, so called because they pass through and develop in the thymus.

Within the thymus, T cells become differentiated. They can be recognised because they carry different markers on the cell surface. These are closely associated with epitopes that react with different monoclonal antibodies. The epitopes are known as 'clusters of differentiation': CD for short. Over 70 CD markers have now been recognised but there are two that occur on very important T cell types. T helper cells, accounting for about 65% of circulating T cells, carry the CD4 antigen and T cytotoxic cells, about 35% of circulating T cells, carry the CD8 marker. All mature T cells carry the CD3 marker. T helper cells interact with B cells to increase antibody production and also help T cytotoxic cells to kill other cells. T cytotoxic cells kill virus-infected cells, tumour cells and 'foreign' cells present in organ or tissue transplants and can interact with B cells to suppress antibody production. In health, cell-mediated immunity requires a balance of CD4 and CD8 cells. If this is disturbed then the affected individual can suffer extreme consequences. In full-blown AIDS, CD4 cells are lost and CD8 cells become dominant. This causes the extreme susceptibility of AIDS patients to opportunistic infections and certain tumours.

Our defences against infection are constantly being probed by potential pathogens and we are constantly mounting a complex response to these attacks. Our responses are usually adequate to protect us from infection. If our immune system overreacts, however, we will suffer as a consequence. Such reactions are described as **hypersensitivity** and these occur on second or subsequent exposure to antigens. There are four major classes of hypersensitivity reaction.

Type I hypersensitivity is an immediate reaction, mediated by IgE. It is also known as **anaphylaxis**. In its most severe form this is a life-threatening condition as smooth muscles in the bronchioles constrict, making breathing difficult. Generalised anaphylaxis can be seen as a reaction to wasp and bee stings and upon second exposure to penicillins. Asthma is a type I hypersensitivity reaction, often precipitated by a respiratory virus infection. Hay fever and the skin rash known as hives are localised type I hypersensitivities. Some virus skin rashes may represent type I hypersensitivity.

Type II hypersensitivity is also known as a cytotoxic reaction. This is typified by transfusion reactions when patients receive blood from someone of a different blood group. Type III hypersensitivity reactions are associated with the formation of immune complexes. Streptococcal infections may lead to kidney damage. This is caused by circulating immune complexes blocking the glomeruli. Type IV hypersensitivity is cell-mediated and involves a delayed T cell response. The skin test used to see if someone has had tuberculosis (TB) is an example of a type IV hypersensitivity reaction. Tuberculin is a protein purified from *Mycobacterium tuberculosis*. This is inoculated into the skin of a person to be tested for exposure to TB. If the individual has not been exposed to TB there will be little reaction to the skin test. If, however, the person has had TB, within eight hours of the skin test the area of inoculation will become markedly reddened and after two days a frank reaction will be seen.

All our defences operate to keep us healthy and free from infection. There are times, however, when ill health is the consequence of other causes that pre-dispose us to infection. Nowhere is this more apparent than when people have to spend time in hospitals. They are, after all, full of sick people. There are many factors that pre-dispose hospital patients to infection. Underlying ill-nesses may weaken our immune defences. Simply lying in bed puts a patient at an unusual angle. This may, for example, cause problems of urine retention. In turn, this makes it more likely that bed-ridden patients will develop a urinary tract infection.

Hospitalised patients are much more likely to suffer breaches of our natural anatomical defences. They are much more likely to have urinary catheters and intravenous devices inserted than people in the general population. It is not unusual for seriously ill patients to require artificial ventilation to assist with breathing. It has been said that every surgical operation is an experiment in bacteriology. All of these breaches increase significantly the risk that the patient affected will acquire an infection. There is also a risk posed by surgical techniques. Clean wounds in areas where there is little or no indigenous flora are unlikely to be contaminated but a significant minority of operations involving the bowel do become infected. Patients hospitalised because of serious burns are also vulnerable to colonisation and infection with bacteria from the hospital environment.

About one hospital patient in every ten acquires an infection as a direct result of their stay in hospital. Such infections are referred to as **nosocomial** infections. This term is derived from the Greek word *nosos*, meaning disease. Specialised nosocomial bacteria often cause these infections. These bacteria are rarely found to cause infection in the population at large but are often seen causing infections in hospitalised patients. Important nosocomial pathogens include pseudomonads and members of the Enterobacteriaceae including *Klebsiella, Citrobacter, Enterobacter* and *Serratia* species, as well as the infamous methicillin-resistant *Staphylococcus aureus*. Frequently they are adapted to the hospital environment and are often resistant to a range of antibiotics. The pattern of antimicrobial susceptibility can be difficult to predict in nosocomial pathogens. This can make treatment of nosocomial infections very difficult. In the case of vancomycin-resistant enterococci, a newly emerged pathogen on high-dependency units, there are no conventional antimicrobial therapies that can be used to control the infections it causes.

Control of nosocomial infection is a full time job. Each hospital has an infection control team. This may seem to be an extravagance but when one considers the number of patients affected by nosocomial infection and remembers the cost of keeping someone in hospital then the cross-infection control team can be seen as an indispensable asset. It is the role of the hospital infection control team to monitor the level of nosocomial infection and, when incidents arise, to respond by identifying the source of the infection and its mode of spread. With this information, it is usually a simple matter to break the cycle of infection.

Often simple measures can be employed to prevent cross-infection from one patient to another. Face masks stop infection from air-borne droplets. Surgical gloves prevent microbes from the surgeon's hands from entering wounds. The use of gloves owes more to chance than to planning; surgeons first wore gloves because phenol solution was sprayed over open wounds during surgery. They wanted to protect their own hands from being burned. Preventing infection in their patients was an added bonus. Sterilisation of surgical instruments and other medical items ensures that pathogens are not accidentally introduced into the patient undergoing medical procedures. There have been incidents where salmonella infections have been spread through inadequately treated endoscopes. Particular care must be taken with the sterilisation of neurological instruments because of the risk of Creutzfeldt–Jakob disease. The agent that causes this infection is particularly resistant to normal sterilisation procedures.

Patients who are suffering from highly contagious infections are often barrier nursed. These techniques have evolved to protect the environment from contamination with a dangerous pathogen. Anything that comes into contact with the patient is considered contaminated and is sterilised before being returned into general use. Personnel are protected by special clothing, gloves and masks. The lessons learned from barrier-nursing techniques can also be used to protect patients who are most vulnerable to infection, such as organ-transplant recipients immediately following the organ graft. In such cases, reverse barrier-nursing techniques are used. This protects the patient from the environmental dangers. The single most important measure in the control of infection, however, is adequate hand washing.

7.2　What are urinary tract infections?

In health, the urinary tract is sterile. For a short distance from its opening the urethra may be colonised by a few bacteria typically found on the skin but these are easily removed by the mechanical flushing of urine. Regular passing of urine helps to maintain the sterility of the urinary tract. This and other innate defences are far from perfect. Urinary tract infections are among the most common of human bacterial infections. They range from relatively trivial episodes of **cystitis** to potentially life-threatening infections of the kidney referred to as **pyelonephritis**.

Cystitis, an infection of the bladder, is trivial only in the sense that it does not lead to permanent damage or death. Its symptoms make any sufferer's life a complete misery during an attack. Pyelonephritis is a kidney infection that is most often caused by an infection ascending the urinary tract from the bladder but may sometimes be caused by the **haematogenous spread** of infection from another site in the body. Patients with pyelonephritis are prone to develop septicaemia. This is a life-threatening, systemic infection associated with microbes circulating in the bloodstream.

Bacteria are the cause most urinary tract infections. Virus or fungal infection of the urinary tract is a rare event and suggests an underlying health problem in the affected patient. In most cases, a single pathogen is isolated from an infected urine sample. Isolation of more than one bacterial type, unless repeated over a course of time, would generally indicate that a specimen was contaminated rather than infected. Because urinary tract infections are so common, diagnostic microbiology laboratories put considerable effort into their diagnosis. Although only single pathogens are implicated in urinary tract infections, their diagnosis is complicated because of the difficulties of obtaining good quality specimens.

7.2.1   What causes urinary tract infections?

Bacteria that are derived from the patient's commensal flora cause most urinary tract infections. The 'coliform' bacteria cause many infections. These are facultative bacteria such as **uropathogenic** strains *Escherichia coli*, found in the human bowel flora. Other bacteria may be found in particular circumstances.

Many more women than men suffer urinary tract infections. In an otherwise fit young man a urinary tract infection is a very unlikely event. In contrast, women frequently suffer a bout of cystitis. Some women endure repeated attacks over a considerable time. The different attack rates of urinary infections between men and women may be explained since many of the pathogens causing urinary tract infections are derived from the bowel. The anatomical differences between men and women mean that pathogens can much more easily enter the female urinary tract from the bowel than they can the male urinary tract.

Anatomical considerations can also explain why more infant boys than girls are prone to urinary tract infections. Male babies are much more likely to suffer congenital abnormalities that disrupt urinary flow than are baby girls. Likewise, the risk of urinary tract infections is considerably increased in older men. The reason for the increased risk is that older men suffer from an enlargement of the prostate gland. This, in turn, leads to a reduction of the normal flow of urine. This increases the likelihood of the sufferer acquiring a urinary tract infection.

Pregnancy brings problems. As a fetus develops, it causes considerable pressure on the soft organs in its mother's abdominal cavity. Pregnant women are at an increased risk of developing a urinary tract infection because of pressure on the bladder. In pregnancy such infections may not be associated with typical symptoms. In the early stages women may even be unaware that they have an infection. They suffer from **asymptomatic bacteriuria**. This condition is, however, potentially very serious since if it is untreated it may lead both to fetal damage and the risk of serious complications for the mother. Consequently, it is usual practice to examine urine samples from women attending antenatal clinics to exclude asymptomatic bacteriuria.

Women who are sexually active are more likely to suffer urinary tract infections than celibates. Those who have recently started engaging in sexual intercourse are at risk of developing 'honeymoon cystitis'. This is caused by *Staphylococcus saprophyticus*, a commensal of the skin in the region around the groin. It gains access to the urinary tract through the mechanical processes of intercourse. Having gained access it then establishes an infection that leads to cystitis.

Regular and complete flushing of the bladder is an important protection against urinary tract infection. Any restriction in the flow of urine, not just those associated with anatomical problems, increases the risk that a person will develop a urinary infection. Neurological problems that affect bladder emptying are associated with an increased risk of urinary infection. These include multiple sclerosis and spina bifida. Stones within the urinary tract, referred to as **calculi**, can also disrupt the urinary flow and hence pre-dispose the sufferer to further infection. Urinary stones are themselves often the result of infection. The causative agent is most likely to be a member of the genus *Proteus*. These bacteria produce a urease enzyme that causes the breakdown of urea to release ammonia. This, in turn, raises the pH of the surroundings. In urine, a raised pH is responsible for the precipitation of inorganic salts that are normally dissolved. These precipitated salts form the focus for the development of a stone.

Table 7.1 *Agents causing urinary tract infections*

Community-acquired infection		Hospital infection	
Escherichia coli	80%	*Escherichia coli*	40%
Coagulase-negative staphylococci	7%	'Other' Gram-negative bacteria	25%
Proteus mirabilis	6%	'Other' Gram-positive bacteria	16%
'Other' Gram-negative bacteria	4%	*Proteus mirabilis*	11%
'Other' Gram-positive bacteria	3%	*Candida albicans*	5%
		Coagulase-negative staphylococci	3%

To assist urinary flow, patients may have a catheter device introduced into the urinary tract. This is a sterile tube passed up the urethra and into the bladder. Its purpose is to assist the flow of urine, which collects in a bag at the other end of the catheter tube. Catheter devices, however, breach our normal anatomical defences and provide an access for bacterial infections. The use of urinary catheters increases the risk of infection and this risk increases dramatically if the catheter must remain in place for a long period. Indeed, patients who have indwelling catheters are not treated if bacteria are found in their urine unless they are suffering from the symptoms of a more generalised infection.

There is a difference between the causes of urinary tract infections suffered by patients in hospital and those in the community. These are summarised in Table 7.1.

The Gram-negative bacteria other than *Escherichia coli* that cause urinary tract infections in hospitalised patients include *Klebsiella* species, *Enterobacter* species, *Citrobacter freundii*, *Serratia marcescens* and *Pseudomonas aeruginosa*. These bacteria are often resistant to a range of antibiotics and are rarely seen causing infection outside the hospital setting.

Bacteria ascending from an infected bladder cause most kidney infections. There are, however, important but rare exceptions to this generalisation. *Staphylococcus aureus* may spread through the bloodstream to cause a localised infection within the kidney that can lead to the development of a renal abscess. Similarly, in patients with tuberculosis, their infection may spread to involve the kidney to cause renal tuberculosis.

Approximately 5% of hospital patients suffer from urinary tract infections caused by the fungus *Candida albicans*. This is a possible cause of urinary infection in patients with diabetes. At one time it was thought that this was simply because diabetics have glucose in their urine and that the fungus was using this as an energy source. Now, however, other factors are also thought to predispose diabetics to infection with *Candida albicans*. In these patients infections are generally confined to the bladder. *Candida albicans* infection is also associated with the presence of indwelling catheters or with surgery on the lower urinary tract. These infections typically resolve once the underlying cause is treated.

7.2.2 What are the symptoms of urinary tract infections?

The symptoms of cystitis include the need to pass urine often, referred to as **frequency** and the overwhelming desire to pass urine: **urgency**. These are often accompanied by **dysuria**: a pain or burning sensation or difficulty in passing urine. Dysuria is also a common feature of infections caused by certain sexually transmissible pathogens. Infected urine is often smelly. It is also likely to appear cloudy. A slight turbidity may be simply caused by the presence of a large number of bacteria. If the cloudiness is more pronounced, this is likely to indicate the presence of **pus** – referred to as **pyuria**. If the urine appears dark in colour this may be because of the presence of blood: **haematuria**. Dark urine is not, however, simply associated with urinary tract infections. Patients with liver disease often pass a dark, frothy urine.

The symptoms of pyelonephritis resemble those of cystitis but also include loin pain and a rise in body temperature: **pyrexia**. Patients with urinary tract infections may suffer considerable abdominal pain and it can be difficult to differentiate urinary infections from acute appendicitis. Examination of a urine sample in patients with suspected appendicitis can prevent unnecessary surgery. Renal tuberculosis is frequently associated with a persistent pyuria that is repeatedly sterile upon routine culture.

7.2.3 How may the diagnostic laboratory assist in the diagnosis of urinary tract infections?

Urinary tract infections are most commonly diagnosed following an examination of a mid-stream specimen of urine: **MSU**. It is important that a good quality specimen is obtained and patients should be given clear instructions to avoid contaminating the specimen. Obtaining a good specimen is, however, easier said than done. The external genitalia should be carefully and thoroughly washed, using soap and water. Antiseptics should not be used since this may affect the subsequent culture results. The patient should then void the first portion of urine to flush out any microbes from around the opening of the urinary tract. The middle portion of the specimen should then be collected, preferably directly into a sterile specimen container provided for the purpose: hence a mid-stream specimen of urine or MSU. Catheter specimens of urine (**CSUs**) are also frequently examined.

Collecting urine specimens from babies poses particular problems. MSU collection depends upon voluntary bladder control. Catheterisation of babies is not, for obvious reasons, a common procedure. One 'solution' used by paediatricians is to fix a bag around the washed external genitalia and to wait for nature to take its course. Bag specimens are, however, almost invariably contaminated with skin and faecal organisms, making accurate diagnosis impossible. To overcome the problems of contamination, in extreme cases supra-pubic aspirates may be performed. In these procedures, a needle is passed just above the pubic bone and a specimen is withdrawn directly from the bladder. This invasive procedure is painful and is only performed when it is important to obtain an accurate diagnosis.

The best time to collect a urine specimen is when the patient first wakes in the morning. Urine will have been held in the bladder for longer than at any other time during the day and, therefore, bacteria will have had a chance to accumulate. This is particularly important when investigating renal tuberculosis. The causative agent, *Mycobacterium tuberculosis*, is slow growing and the number of bacteria in the sample is generally low. In the diagnosis of renal tuberculosis it is customary to examine three early morning specimens of urine collected on consecutive days. This increases the chances of obtaining a positive culture.

If a urine specimen cannot be examined straight away, steps must be taken to prevent bacterial overgrowth. Urine is a good medium in which to grow bacteria and if a specimen is left at room temperature the number of bacteria in the sample will increase dramatically. This will affect the interpretation of culture results. One way of preventing multiplication of bacteria in urine is to refrigerate samples until they can be processed in the laboratory. This is not a practical solution for specimens that are taken at the request of a General Practitioner and that may need to be transported several miles to the local laboratory. An alternative method is to include boric acid in the specimen jar. This is bacteriostatic and will, therefore, prevent multiplication of bacteria in the sample. Once the specimen is plated onto culture medium, however, the boric acid is diluted into the medium and no longer prevents growth of any bacteria in the sample.

Upon receipt of the specimen by the laboratory, its appearance may suggest an infection. A smelly or cloudy urine suggests infection but other factors may lead to these appearances. More important is the microscopic examination of urine. Besides bacterial cells, the microscopic examination of a wet preparation of urine may reveal leukocytes, red blood cells and 'casts'. These are proteinaceous deposits formed within a diseased kidney. Hyaline casts are clear and made up principally of protein but casts may be granular if they have crystals stuck to them. Casts may also be associated with either leukocytes or red blood cells. If large, flat epithelial cells are seen in a urine sample then this is taken as indication that the specimen is contaminated. This is because such epithelial cells are not found in the urinary tract but are associated with the skin. If there is microscopic evidence of infection, the urine is plated out and an antibiotic sensitivity test is performed directly on the sample. This will save a day in reporting the results.

A limited range of bacteria causes the vast majority of urinary tract infections. Nearly all are capable of growth overnight when plated onto a medium such as CLED agar. This is a non-inhibitory medium widely used for urine microbiology. It is deficient in electrolytes and this prevents swarming of *Proteus* species It also contains lactose and a pH indicator. Those bacteria that can ferment lactose to produce acid appear as yellow colonies whereas the non-lactose fermenters grow into blue colonies. Coliform bacilli are the most common cause of urinary tract infections and generally no attempt is made to identify these further. If this were done, most would be *Escherichia coli*.

Because of the problem of contamination of urine samples, a semi-quantitative culture is typically performed. A standard volume of urine is plated and the number of colonies that it yields is estimated. For a mid-stream specimen of urine, if 1 microlitre is plated, more than 100 colonies of a *single* colony type is considered to indicate a significant urinary tract infection. This is equivalent to 10^5 colony forming units per millilitre of urine. When examining suprapubic aspirates, much lower counts are considered to be significant since the collection procedure is designed to minimise specimen contamination.

Growth of more than one colony type from a urine sample, unless reliably repeated on a number of occasions, is taken as evidence that the sample is contaminated. In such instances, a repeat specimen should be examined if symptoms have persisted. Catheter specimens frequently yield mixed cultures, particularly if the catheter has been in place for more than a few days. Sometimes the number of bacteria in catheter specimens can be very high. If, however, the sample shows a mixed growth then antibiotic therapy should be avoided unless the patient is showing signs of generalised infection such as a raised temperature.

It is not uncommon for patients to have microscopic evidence of infection and yet for the culture to fail to grow a pathogen. This is a particular problem when examining specimens submitted by General Practitioners. A common explanation is that the patient in question suffers from repeated urinary tract infections. After receiving antibiotics to treat a previous episode, the patient stops taking medication when the symptoms subside. The drug is left in the medicine cabinet where they may well lose potency and when the symptoms return the patient starts taking the course again. Up to 25% of urine samples submitted by General Practitioners contain identifiable antimicrobial substances.

It is only in exceptional circumstances that culture-negative urinary tract infections are caused by fastidious organisms, although these do require specialised investigation. One important example is the diagnosis of renal tuberculosis. Early morning specimens are collected on three consecutive days and are plated on a suitable medium such as Lowenstein Jensen medium. This is then incubated for up to eight weeks to grow *Mycobacterium tuberculosis*. Special stains for acid alcohol-fast bacilli are not particularly helpful in the diagnosis of renal tuberculosis since many healthy people carry *Mycobacterium smegmatis*, a commensal bacterium. Its microscopic appearance is easily confused with the pathogen *Mycobacterium tuberculosis*, thus confusing the diagnosis.

7.3 What causes sexually transmissible diseases?

Many of the pathogens that cause venereal infections, sexually transmissible diseases, are vulnerable to the external stresses of temperature change and desiccation. It is for this reason that they have evolved to spread from one person to another through intimate body contact. During sexual intercourse, these pathogens may spread from one individual to another without coming into contact with a hostile external environment. Sometimes, however, the pathogens associated with venereal infections can cause infections by other means. The agents of syphilis and AIDS, for example, can be transmitted through the transfusion of infected blood. It is for this reason that we use the term 'sexually transmissible', indicating an element of doubt, rather than 'sexually transmitted', a phrase that implies a degree of certainty. Viruses, bacteria and fungal infections together with arthropod infestations have all evolved to spread from person to person through sexual contact. A list of some of the important sexually transmissible diseases and their causes is given in Table 7.2.

A major factor in the spread of sexually transmissible disease is promiscu-

Table 7.2 *Agents causing sexually transmissible diseases*

Sexually transmissible disease	Cause
AIDS	Human immunodeficiency virus (HIV)
Syphilis	*Treponema pallidum* (a bacterium)
Gonorrhoea	*Neisseria gonorrhoeae* (a bacterium)
Non-specific urethritis	*Chlamydia trachomatis* (a bacterium)
Lymphogranuloma venereum	*Chlamydia trachomatis* (a bacterium)
Chancroid	*Haemophilus ducreyi* (a bacterium)
Candidosis (thrush)	*Candida albicans* (a fungus)
Trichomoniasis	*Trichomonas vaginalis* (a protozoan)
Genital herpes	Herpes simplex virus
Condylomata accuminata (genital warts)	Papilloma viruses
Hepatitis B and C	Hepatitis viruses
Pubic lice	*Phthirus pubis* (an arthropod)
Scabies	*Sarcoptes scabiei* (an arthropod)

ity, particularly if the individual involved has sexual relationships with several people over a relatively short period. Often promiscuity is linked to poverty but the complex nature of human sexuality makes it dangerous to generalise too much about risk factors for the acquisition of a sexually transmissible infection. Certainly there are high-risk groups who are more likely to acquire these infections, but there is probably no such thing as a no-risk group.

From the point of view of sexually transmissible pathogens, an important strategy for their continued survival is that they frequently cause asymptomatic or sub-clinical infections. If this were not the case, then the diseases they cause could be eradicated. This would be achieved by the simple measure of preventing the infected individual from engaging in sexual intercourse until cured of the infection. The continuing high incidence of sexually transmissible infection shows this is not the case.

For many sexually transmissible diseases the infection may have a latent persistent phase during which infectious particles can be shed and passed to other individuals. Such is the case with infection by HIV, the cause of AIDS. The importance of latency is graphically illustrated by herpes simplex infection. Given the nature of the lesions caused by genital herpes infection, it is unlikely that someone who is suffering from an active episode of genital herpes would willingly engage in intercourse. The lesions are too painful. Herpes simplex infection relies upon people who have a latent infection shedding infectious virus during the asymptomatic phase of the disease. Alternatively, there is a large pool of infected individuals who show no signs of disease. Such individuals may have a truly asymptomatic infection, such as many women who are infected with *Neisseria gonorrhoeae*. The use of condoms can severely limit the spread of venereal infections, particularly when used with non-penetrative sexual practices. Although other barrier methods of contraception may successfully prevent pregnancy, they are not effective in preventing venereal infection and should not be used with this aim in mind.

Sexually transmissible diseases are not confined to infections of the genital tract. Many sexually transmissible pathogens can cause lesions of the throat or the rectum besides the genital tract. The site of infection often reflects the sexual practices engaged in by the infected partners. Some pathogens cause a more generalised infection. This may be as a rare complication of infections such as gonorrhoea or may be the typical presentation of diseases such as syphilis and AIDS.

7.3.1   Acquired immunodeficiency syndrome (AIDS)

It is probable that AIDS originated in Central Africa. With the ease of global travel, it spread to the developed world during the late 1970s. It was recognised shortly after this in the USA. It then rapidly spread around the world. AIDS has a mortality approaching 100%. The syndrome was first described in 1981 and the causative virus was isolated in 1983. The first case clusters to be reported occurred in American cities, most notably San Francisco and New York. Other clusters were soon reported across North America and in Europe. AIDS is now recognised as a pandemic disease and is transmitted in every country.

AIDS is a particular problem in the developing countries. In the so-called AIDS belt of sub-Saharan Africa, up to 10% of rural dwellers and about 25% of city dwellers are currently infected with HIV. The pattern of AIDS in Africa is, however, unstable and in some rural areas the number of people infected with HIV can approach levels more typically seen in urban areas. It is estimated that in 1997, 14 million people in Africa were infected with this virus. At particular risk are the populations of Kenya, Malawi, Rwanda, Tanzania, Uganda, Zambia and Zimbabwe. On average, about 10% of women attending ante-natal clinics in these countries are HIV positive. They were mostly infected as teenagers and through heterosexual intercourse. Yet there is hope. In certain large African cities, prevalence of HIV infection remains low. It is to be hoped that investigation of the reasons for this low incidence will contribute to more effective control measures, which could be applied in areas with a higher incidence of disease.

There is also a large problem with HIV infection in Asia. This is not confined to cities such as Bangkok, notorious for its sex trade. In India, over 50% of sex workers in Bombay are currently infected with HIV. Other areas of Asia, notably Vietnam and Cambodia, are currently experiencing an alarming rise in the number of people infected with HIV. The virus is also spreading rapidly through the populations of Central and Eastern Europe. HIV infection is a crisis of truly global proportion.

HIV causes a persistent infection and once people acquire this virus they remain infectious for life. Infection with HIV ultimately leads to a syndrome characterised by a deficiency in cell-mediated immunity. As a consequence, patients suffer from a spectrum of opportunistic infections caused by microbes that do not typically cause problems in immunocompetent people. The virus infects cells that carry the CD4 antigen. These include T helper cells, monocytes and dendritic cells. Loss of functioning T helper cells leads to immunosuppression in an infected person and this loss is irreversible if untreated in AIDS. Because cells circulating in the bloodstream are infected with HIV, blood and blood products are a major vector for the spread of AIDS. Small amounts of virus may be found in semen and also in breast milk of infected women.

During the early days of the AIDS pandemic, haemophiliacs were particularly vulnerable to AIDS because of their dependence on factor VIII. They lack this blood clotting factor and are given transfusions of factor VIII collected from pooled human blood. The American practice of paying blood donors made this problem worse, since many donors sold blood to feed a drug habit that included intravenous injection. Often addicts share dirty needles and this increases the risk of transmitting such diseases as AIDS, hepatitis B and hepatitis C.

The primary infection with HIV often resembles glandular fever. Patients appear listless, lack an appetite and have swollen lymph glands throughout the body. During this period, the number of virus particles circulating in the blood is relatively high. As the glandular fever-like illness resolves, the infected person appears perfectly healthy and the number of circulating virus particles in the person's blood drops dramatically. The disease then enters a **quiescent** phase.

Patients remain apparently well for an unpredictable period. This period is probably shorter for people in the Third World than for those in developed countries. As a result of triggers that are not well understood at present, ultimately the virus load in the blood rises again. Consequently, there is a constant battle between the host immune system and the replicating virus, which undergoes a number of antigenic changes in the HIV-positive individual. Such people are either rapid, median or slow progressors to full-blown AIDS. Comparative research on these individuals will help to throw light on this infection and the possibility of developing effective controls to prevent the current inexorable progress towards disease and death.

The increase in virus load coincides with the patient suffering from the first opportunistic infections. This also marks the onset of the AIDS-related complex (ARC). Patients with the AIDS-related complex typically suffer weight loss, fever, diarrhoea and swollen glands. They may also suffer from oral **candidosis**, commonly known as thrush. Women with AIDS also frequently develop candida vaginitis, also referred to as vaginal thrush.

As the disease becomes established, the virus load rises further and remains high through the transition to full-blown AIDS. Measuring the virus load in the blood is an important tool in monitoring the progress of people infected with HIV and in following the effectiveness of drug therapy during full-blown AIDS. People with full-blown AIDS suffer from a variety of opportunistic infections. These include *Pneumocystis carinii* pneumonia and persistent cryptosporidial diarrhoea. Patients are also likely to suffer from tuberculosis or similar infections caused by opportunistic mycobacteria, especially those of the *Mycobacterium avium-intracellulare* complex.

There is perhaps a grim irony in this. Towards the end of the nineteenth century, tuberculosis claimed the lives of influential artists such as Frederick Chopin. Consumption, as tuberculosis was once known, was a disease that became almost fashionable. Its victims had a pale and interesting look, very much in vogue in Victorian times. It had a tremendous impact on literature and the arts. Violetta, the heroine of La Traviata, dies of tuberculosis in the final act of Verdi's opera based on *La Dame aux Camélias*, a novel by Alexandre Dumas *fils*. In the twentieth century, AIDS seems to have become almost a modern day equivalent, claiming the lives of popular artists.

Cryptococcal meningitis is rare in otherwise healthy individuals but is not uncommon in patients with full-blown AIDS. *Candida* species are also a problem for AIDS patients. These are associated particularly with lesions in the mouth and oesophagus and are notoriously hard to treat. It was thought that this was because strains of *Candida albicans* in AIDS patients were particularly resistant to chemotherapy. Recent work in Ireland has shown that AIDS sufferers are quickly colonised by a different species: *Candida dubliniensis*. This is intrinsically more resistant to antifungal drugs than *Candida albicans*.

Viruses can also cause a problem for AIDS patients, who commonly suffer from cytomegalovirus pneumonia. Most AIDS patients also develop neurological problems as the disease progresses. This is often the result of the reactivation of latent or persistent viruses. Herpes simplex virus may reactivate to cause **encephalitis**, as may varicella zoster virus. The papovavirus JC can also reactivate, causing a fatal **demyelinating** disease, progressive multifocal leukoencephalopathy (PML).

Patients with AIDS may develop a disfiguring and potentially fatal cancer, **Kaposi's sarcoma**. Between 20 and 25% of patients develop this tumour. Until the onset of the AIDS pandemic, this was a very rare condition typically only occurring in old men of Jewish or Mediterranean extraction. Indeed, it was the sudden rise in incidence of Kaposi's sarcoma along with the unusually high incidence of opportunistic infections in young, homosexual men that first alerted the world to the emergence of AIDS. There is mounting evidence to show that the development of Kaposi's sarcoma results from the reactivation of a latent virus infection caused by a herpes-like virus, possibly human herpesvirus 8.

Treatment of AIDS is difficult and depends upon giving a cocktail of drugs aimed at helping to prevent the progress of the underlying HIV infection and in trying to control the symptoms of full-blown AIDS. The mainstay of treatment is the nucleotide analogue zidovudine (AZT) and the antiretroviral protease inhibitors. It is important that people who are HIV positive are regularly monitored to assess their level of T cells and the virus load; both are important markers in the progress of the disease. Results of such tests can help to guide the optimal therapy regimen for individuals.

Transmission of HIV infection is most often through sexual intercourse. During the early days of the AIDS pandemic it was thought that the disease was confined largely to homosexual males. This led to the descriptions of 'Gay Plague'. Now we realise that heterosexual intercourse is an efficient means of spreading HIV infection and hence AIDS. The infection may also be transmitted through contact with infected blood or blood products. Today, this happens most often when intravenous drug users share infected needles. To complicate matters, some addicts turn to prostitution to get enough money to feed their habit. Consequently, they become a threat to their sex clients, furthering the spread of HIV infection. Out-reach initiatives have shown that needle exchange programmes have a dramatic effect in reducing the incidence and spread of HIV infection among intravenous drug injectors. HIV-infected mothers may also pass the infection to their unborn babies. About 20% of babies born to infected mothers are themselves infected with HIV, or become so shortly after birth.

Laboratory diagnosis of AIDS relies upon demonstrating the presence of specific antibodies in the blood of infected people. Initially, **ELISA** (enzyme-linked immunosorbent assay) tests are used to screen for infection. Because of the very serious nature of the diagnosis, however, **Western blotting** (using antibodies to identify proteins separated by gel electrophoresis) and immunofluorescence testing are used to confirm positive results.

7.3.2 Syphilis

Syphilis is now uncommon in developed countries. It remains important, however, because of the high risk of patients developing serious complications from their infection. There is also a significant risk that babies born to infected mothers will suffer from congenital syphilis. The serious risks associated with syphilis are perhaps well illustrated by the pre-antibiotic therapy for syphilis. Before the development of antibiotics that could treat syphilis, it was considered an ethical treatment to infect syphilis victims with malaria. The potentially fatal consequences of this treatment were judged to be better for the patient than was the risk of developing tertiary syphilis.

Syphilis was first recognised in the sixteenth century when the sailors who accompanied Christopher Columbus introduced the disease on their return to Europe from the New World. It then spread around the world. During the Victorian era at the end of the nineteenth century, the disease was in decline but numbers increased dramatically during and after the First World War. At the time of the Second World War, simple diagnostic tests were developed and penicillin became widely available. These two events caused a dramatic reduction in the number of cases of syphilis seen in industrialised countries. The relaxation of sexual morality in the 1960s caused another increase in numbers that progressed until the AIDS scare of the mid 1980s. Although syphilis is not common in developed countries, it is still prevalent elsewhere in the world. Recently, there has been a cluster of cases amongst British executives returning from business trips to Central Europe, for example.

The causative agent of syphilis is *Treponema pallidum* a slender spirochaete that is difficult to visualise using conventional microscopy. It cannot be grown in artificial culture, although it may be maintained in experimentally infected laboratory animals. It is extremely sensitive to heat. It is this property that led to the introduction of malaria treatment, since patients with malaria suffer regular, periodic and predictable bouts of fever. The bacterium is most often transmitted during sexual intercourse but can also be passed on through infected blood transfusions.

Symptoms of the disease fall into three separate stages: primary, secondary and tertiary syphilis. Not all patients progress through all three stages. Following initial infection, patients develop an ulcer at the site of infection.

This is known as a **hard chancre**. The incubation period is from two to ten weeks following initial contact. The appearance of the chancre is often associated with swollen lymph glands. The chancre appears at the site of infection, typically on the genitalia. Chancres may, however, appear on the lips or in the oral cavity if the patient has engage in oral sex. Syphilitic chancres are painless and heal spontaneously and so are often ignored.

Following regression of the chancre, an infected person appears healthy until the onset of secondary syphilis, one to three months later. This is marked by a flu-like illness. Patients complain of headaches, **myalgia** (muscle pain) and general malaise and anorexia. The secondary stage of syphilis is also characterised by the appearance of a generalised rash that does not itch. Patients may also develop snail-track ulcers in the oral cavity. Another common symptom of secondary syphilis is patchy baldness. Warts may appear around the mouth and the anus. Symptoms last from two to six weeks. Again, the lesions heal and the disease goes into remission.

After a variable period in remission, about one-third of patients with syphilis develop the tertiary disease. The period of remission is highly unpredictable but is seldom less than three years and often not more than 30 years in duration. The symptoms of tertiary syphilis affect multiple organs and are not easily defined. There are, however, some important features that occur commonly. Patients with tertiary syphilis frequently suffer severe, widespread ulceration of the skin and mucosal surfaces of the body. The ulcers are known as **gumma**. Tertiary syphilis also leads to a weakening of blood vessels. There is a significant risk that large arteries, in particular the aorta, will weaken to such an extent that the wall ruptures, leaving the patient to die from haemorrhage. Another clinical condition associated with tertiary syphilis is **tabes dorsalis**. This is caused by nerve damage and is characterised by shooting pains in the legs, uncoordinated movement and loss of nervous reflex reactions. Neurosyphilis, once referred to as generalised paralysis of the insane, is characterised by loss of recent memory progressing to dementia, convulsions, loss of bladder and anal sphincter control, euphoria and delusions of grandeur. The clinical manifestations of tertiary syphilis are not directly a result of the infectious process. Rather, they are caused by an immunological hypersensitivity reactions.

The long-term effects of syphilis have been studied in a highly controversial and unethical experiment that ran from 1932 to 1972. A cohort of 600 African-American males was identified from a low-income population in Tuskegee, Alabama. Of these, 400 were, unknown to them, infected with *Treponema pallidum*. Even though effective treatment became widely available during the early 1950s patients in the infected group were actively prevented from receiving antibiotics. The US Department of Health, Education and Welfare only stopped this experiment when details were leaked and the government became politically embarrassed by the existence of the programme.

Although *Treponema pallidum* is difficult to visualise under standard light-microscopic techniques, it can be seen when heavily infected material is viewed using dark-ground microscopy. This can be used as an initial screen in the diagnosis of syphilis but is not a reliable method. Spirochaetes can rapidly drop to sub-detectable levels and within just 30 minutes of taking penicillin spirochaetes may disappear from infected material when visualised microscopically. *Treponema pallidum* cannot be grown in artificial culture, except in experimental animals. Consequently, the diagnosis of syphilis currently relies upon serological testing.

Serological tests for syphilis may be non-specific or specific, although the so-called specific tests are specific only for the genus *Treponema*. Other non-venereal infections caused by treponemes will give positive results when specific tests for syphilis are used. These include the tropical infections pinta and yaws. The VDRL test, named after the Venereal Disease Reference Laboratory, is a non-specific test for syphilis. Although it is non-specific, this is a very useful test because a positive VDRL test becomes negative when a person is successfully treated. There are two widely used specific tests for syphilis: the TPHA test and the FTA(abs) test (Fig. 7.2). Once a patient has a positive response to these tests they remain positive for life, even if the disease has successfully been cured.

The VDRL test demonstrates the presence of **reagin**. Its precise nature has been debated since the early days of syphilis serology but reagin is the complement-fixing substance that reacts with a lipid extract found in the blood of patients with syphilis. The VDRL test is a flocculation test. Antibodies in the serum of an infected patient will cause flocculation of the cardiolipin antigen. The antigen for the VDRL test is made by an alcoholic extraction of lipid from beef heart. The test has been enhanced by the incorporation of carbon particles in the reaction mixture. These make the clumping seen in positive tests easier to visualise.

In the TPHA test, short for *Treponema pallidum* haem-agglutination test, treponemal antibodies are attached to the surface of avian red blood cells. The treated cells will agglutinate in the presence of anti-treponemal antibodies in a positive serum. The test is performed in a microtitre tray and the agglutinated cells form a mat that spreads across the bottom of the well. In contrast, negative serum does not cause the cells to agglutinate and so the red blood cells fall to the centre of the well, where they form a tight button. Untreated red blood cells are used as a control with each serum to be tested, to detect

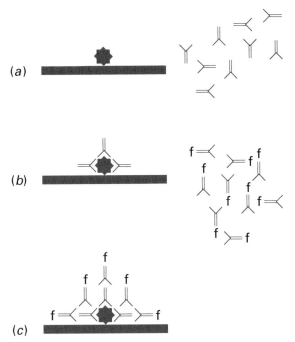

Fig. 7.2. The FTA(abs) test for syphilis. (*a*) Cells of *Treponema pallidum* are attached to a microscope slide and are flooded with serum to be tested. (*b*) If the serum is positive it will contain antibodies that can attach to the treponemes. (*c*) The slide is washed, then treated with fluorescently tagged anti-human antibodies. These will attach to the antibodies from the test serum that have stuck onto the treponemes fixed onto the slide. The slide is washed and viewed in a fluorescence microscope. In a positive sample, the fixed bacterial cells will shine brightly because of the antibody sandwich. Negative sera do not contain antibodies with which the tagged anti-human antibodies can react and these will be washed away. Therefore, the bacteria in negative tests cannot fluoresce.

false-positive reactions. This control shows that agglutination results from the presence of anti-treponemal antibodies, rather than because of an innate agglutination reaction with that serum sample.

In the fluorescent treponemal antibody absorption test, FTA(abs), anti-treponemal antibodies from a positive serum sample attach to treponemes that have previously been fixed onto a microscope slide. These are then detected by flooding the slide with fluorescently labelled anti-human antiserum. If the serum under test is positive, then the fluorescently tagged immunoglobulin will bind to the anti-treponemal antibodies that are themselves bound to the bacterial cells attached to the microscope slide. This forms a type of molecular sandwich. When viewed in a fluorescence microscope, positive samples will be easily detected since the bacterial cells on the slide will fluoresce brightly. Negative samples do not contain anti-treponemal antibodies so there is nothing for the anti-human antiserum to stick to and, therefore, the bacteria will not appear fluorescent. By using either anti-human IgM or anti-human IgG, the time since infection may be estimated. Upon initial infection, IgM is produced as a first response and only later do IgG antibodies appear. The FTA(abs) test is the first serological test for syphilis to become positive following infection.

In all serological tests for syphilis, there is a small chance of obtaining a false-positive reaction. It is, therefore, important that any positive reaction is confirmed by at least one other test. Patients with autoimmune diseases, in particular **systemic lupus erythematosus** (SLE), are prone to have false-positive TPHA and FTA(abs) tests, as are intravenous drug abusers. VDRL false-positive reactions are occasionally seen during pregnancy and in patients with malaria and leprosy. Patients with particular autoimmune diseases may also yield false-positive VDRL reactions.

Because there is an effective array of diagnostic tests for syphilis, control of this infection can be achieved. Blood donors are routinely screened for the disease to prevent transmission in infected blood transfusions. Other at-risk groups can be screened. Because of the risk and serious consequences of congenital syphilis, pregnant women are routinely screened in ante-natal clinics. Once a positive case is confirmed, careful and confidential contact tracing helps to limit the spread of this infection.

There have been many attempts to find a cure for syphilis over the years. It used to be said that 'a night with Venus would lead to a lifetime with Mercury'. This is an allusion to the use of mercury salts to try to cure the disease. Other metal ions, including arsenic, have also been used to try and cure this disease. Fortunately, *Treponema pallidum* is very sensitive to penicillin and this is now the recommended treatment. For patients who are allergic to penicillin, erythromycin provides an effective alternative. Patients who are given antibiotics can suffer a severe toxic reaction known as the Jarisch–Herxheimer reaction. It is caused as the dying bacteria release toxins that then elicit a clinical response. Patients suffer fever, rapid pulse and breathing and have low blood pressure. This may be avoided if patients are given a corticosteroid together with the antibiotic.

7.3.3 Gonorrhoea

The bacterium that causes gonorrhoea, *Neisseria gonorrhoeae*, is exquisitely sensitive to drying and for this reason its mode of transmission relies upon sexual intercourse. Of the women who are infected with *Neisseria gonorrhoeae*, only about 20% show signs of the disease. The remainder are asymptomatic carriers. In contrast, only about 10% of the men who are infected remain free of symptoms. Asymptomatic carriers act as the reservoir of infection for *Neisseria gonorrhoeae*, thus perpetuating the disease. They are, however, at risk of developing non-genital complications because of their infection. These complications may be life-threatening.

Uncomplicated genital infections are the most common symptomatic *Neisseria gonorrhoeae* infection. From two to seven days following infection, patients develop a discharge of pus from the urethra and may have difficulty in passing urine. Although patients may need to pass urine more frequently than usual, it is often accompanied by an intense burning sensation. In men whose prostate gland becomes infected, passage of urine may be restricted, or even blocked. Although genital infections are the most common, certain sexual practices also lead to infections of the rectum or the throat. The symptoms of genital gonorrhoea are indistinguishable from non-gonococcal urethritis, most often caused by *Chlamydia trachomatis*. Consequently, microscopy, culture and antimicrobial sensitivity testing play an important part in the diagnosis of gonorrhoea. The presence of Gram-negative intracellular diplococci inside **polymorphonuclear leukocytes** in a urethral discharge is sufficient evidence to make a provisional diagnosis of gonorrhoea. To culture *Neisseria gonorrhoeae*, clinical specimens are plated onto a rich medium and are incubated for up to 48 hours. Urethral, cervical and high vaginal swabs are plated onto VPAT agar. This is a rich, blood-based medium containing a cocktail of antibiotics. These prevent the overgrowth of commensal microbes. Vancomycin is active against Gram-positive bacteria; polymyxin kills most but not all Gram-negative facultative bacilli. Amphotericin B is included to inhibit fungal growth and trimethoprim is a broad-spectrum antibacterial used to plug any gaps in the antimicrobial armoury. Bacteria of the genus *Neisseria* are resistant to all of these agents and so their colonies can grow on VPAT agar. Cultures are incubated under 5–10% carbon dioxide. Rectal and throat swabs are similarly treated, if appropriate.

Among the complications of gonorrhoea are pelvic inflammatory disease and septic **arthritis**. In women with untreated or asymptomatic gonorrhoea, the causative bacterium may ascend through the uterus to infect the fallopian tubes. The damage caused may lead to partial or complete blockage, causing infertility. In partially blocked fallopian tubes, there is a significant risk that the woman will suffer an ectopic pregnancy. This is where the fetus develops within the fallopian tube rather than in the uterus and it is a life-threatening condition. Men who leave their gonococcal infections untreated may suffer an ascending infection that damages the testes, again leading to infertility problems. Babies born to infected mothers may acquire eye infection as they pass through the birth canal. This can lead to a condition known as **ophthalmia neonatorum** and can lead to blindness if untreated. Occasionally, *Neisseria gonorrhoeae* can gain access to the bloodstream. There it can travel around the body to cause infections at other sites. This can lead to infection of a joint. The synovial fluid becomes filled with pus and the affected joint becomes painful, red and swollen. More rarely, *Neisseria gonorrhoeae* may cause endocarditis.

The preferred treatment for gonorrhoea is penicillin. For many years *Neisseria gonorrhoeae* was sensitive to penicillin but during the mid-1970s, penicillin-resistant strains were first recognised. They first arose in two independent evolutionary events, in Africa and in southeast Asia, from where they rapidly spread around the world. It is not unusual for people with gonorrhoea to suffer infections with other pathogens, especially *Chlamydia trachomatis*. This is an obligate intracellular parasite that lacks a peptidoglycan cell wall. Consequently, penicillin is unable to treat infections caused by this bacterium. In cases of double infection, tetracycline is the treatment of choice. In the mid-1980s, however, tetracycline-resistant strains of *Neisseria gonorrhoeae* evolved. These have also spread around the world. Very shortly after the appearance of tetracycline-resistant *Neisseria gonorrhoeae*, strains resistant to both tetracycline and penicillin appeared. Fortunately, these are not common and such infections can be treated with an extended-spectrum cephalosporin or an aminoglycoside.

7.3.4 Non-specific urethritis and other bacterial infections

An alternative name for non-specific urethritis is non-gonococcal urethritis and its symptoms are indistinguishable from gonorrhoea. It is caused by *Chlamydia trachomatis*. *Mycoplasma hominis* and *Ureaplasma urealyticum* have been implicated as occasional causes of non-specific urethritis but evidence for their role in the disease is not overwhelming. *Chlamydia trachomatis* is a very simple bacterium that is an obligate intracellular parasite. As with gonorrhoea, many chlamydial infections are asymptomatic. Women are particularly likely to have an asymptomatic infection. Eye infections are a common complication of non-specific urethritis. These are commonly caused through auto-inoculation. After touching infected material from the genital area, patients then transfer the infection by rubbing around the eyes. Because chlamydia are intracellular and because they lack a cell wall, the treatment of choice for chlamydial infection is tetracycline.

Chlamydia can only be propagated in the laboratory using tissue culture techniques: McCoy cells are commonly employed. As a diagnostic method, this is very insensitive and many false-negative reports result from the sole reliance upon this method of diagnosis. To overcome the problem of sensitivity, antigen detection tests have been developed and fluorescent antibody tests are often used in the laboratory diagnosis of chlamydial infection. Currently, a PCR test is becoming established as a highly sensitive method for the diagnosis of infections caused by chlamydia.

Reiter's syndrome is an uncommon complication of chlamydial infection. It is characterised by **conjunctivitis** and arthritis, most often affecting the lower back and legs, as well as urethritis. It most often affects young men with the HLA-B27 tissue type and it is thought to be mediated through an immune mechanism. It usually manifests itself one to four weeks following infection and although it spontaneously resolves itself, recurrences are seen in about 25% of patients. Reiter's syndrome is also associated with certain gastrointestinal infections, especially those caused by *Shigella* species, *Yersinia* species and campylobacters.

Certain strains of *Chlamydia trachomatis*, particularly serovars L1, L2 and L3, are associated with a more serious condition: **lymphogranuloma venereum**. This is a condition that is more common in Africa, Asia and South America than in Europe and North America, and it particularly affects homosexual men. One to four weeks following inoculation, the site of infection becomes ulcerated. Patients also suffer headaches and myalgia. As the infection develops, the lymph glands in the groin become very swollen. Chlamydia may spread to the rectum where they cause a painful inflammation referred to as **proctitis**.

In women, *Chlamydia trachomatis* commonly infects the cervix and the urethra. It is also, however, a common cause of uterine infection, known as **endometriosis**, and **salpingitis**, which is an infection of the fallopian tubes. Approximately one in ten cases of cervical infection with *Chlamydia trachomatis* can lead to pelvic inflammatory disease and its complications. The risk of complications increases dramatically if women suffer repeated episodes of salpingitis. There is a 10% risk of infertility after the first episode of salpingitis but this rises to 30% after the second occurrence and to 50% after the third attack.

Pelvic inflammatory disease may be caused by pathogens other than *Chlamydia trachomatis* and *Neisseria gonorrhoeae*. These include *Mycoplasma hominis* and a variety of anaerobic bacteria including *Actinomyces* species. Actinomycete infections were especially associated with particular old designs of intra-uterine contraceptive devices. Anaerobic incubation of samples inoculated onto fresh blood agar is maintained for five days to aid the detection of *Actinomyces israelii*. Severe infections of the uterus and pelvic organs can be caused by *Clostridium perfringens* or by *Streptococcus pyogenes*. Such infections represent complications of gynaecological surgery, particularly 'amateur' abortions.

Chancroid is caused by the fastidious Gram-negative bacillus *Haemophilus ducreyi*. This bacterium can be difficult to culture and laboratory diagnosis may depend upon the observation of short Gram-negative bacilli or cocco-bacilli in clinical specimens. Patients with chancroid suffer soft, painful ulcers at the site of infection. These are known as chancres and resemble the lesions of genital herpes infections. The recommended treatment for chancroid is erythromycin.

7.3.5 Candidosis (thrush)

The **dimorphic fungus** *Candida albicans* lives as a commensal organism in the vagina of some women. In its commensal state, *Candida albicans* is most often present in its yeast form. For reasons that are not clearly understood, the commensal status can become disturbed. The yeast reverts to a mycelial form and the fungus overgrows, causing a 'cheesy' vaginal discharge. This may be associated with urethritis and dysuria. Use of oral contraception was at one time thought to be a pre-disposing factor for the overgrowth of *Candida albicans*, but this is no longer certain. Broad-spectrum antibiotics do, however, predispose women to vaginal candidosis. Candida infection is not confined to women. About 10% of male partners develop white plaques on the penis, known as candida **balanitis**. When an individual develops clinical candidosis then asymptomatic sexual partners must in some cases also be treated if the condition is not to recur in the short term. Treatment for candidosis can be either oral or topical. Clotrimazole pessaries and creams are a popular method of treating candidosis.

7.3.6 Trichomoniasis

Despite its name, *Trichomonas vaginalis* is an obligate anaerobic flagellate protozoan that can colonise the urethra of men. In the male urethra, this organism does not cause a problem. Men who are colonised are, however, a reservoir of infection for women, who suffer vaginal irritation and a copious, foul-smelling vaginal discharge when infected. Microscopic examination of infected material allows the observation of motile, flagellate trophozoites. Trichomoniasis can be effectively treated using metronidazole.

7.3.7 Genital herpes infections

Herpes simplex virus can be divided into two types, 1 and 2, differentiated by restriction endonuclease digestion patterns of their DNA and by their antigenic structure. Differentiation is not routinely undertaken in diagnostic laboratories. Herpes simplex type 1 causes cold sores, whereas herpes simplex type 2 is associated with genital infections. Oro-genital sexual practices mean that herpes simplex type 1 is often isolated from genital lesions but it is unusual that herpes simplex type 2 causes oral lesions. Consequently, genital herpes infections were once thought to be associated with herpes simplex type 2. Now both herpes simplex type 1 and type 2 viruses can be isolated from genital lesions, with the ratio of type 1 to type 2 infections depending upon geographic distributions.

The first lesions appear between three and seven days following infection. The first evidence of infection is blistering and the resulting vesicles develop into painful ulcers. Lymph glands in the groin then become swollen and tender and patients may suffer more generalised flu-like symptoms. Although the lesions heal within two to three weeks, the virus migrates to the dorsal root ganglion where it establishes a latent infection. Periodically this latent infection may reactivate and the virus will travel back along the nerves to cause lesions on the genitalia again. Use of the antiherpes drug aciclovir may shorten the period during which patients suffer clinical symptoms but it does not eradicate latent virus.

Babies born to mothers with active genital herpes are at risk of developing a disseminated and overwhelming herpes infection. This may be avoided if the infant is delivered by caesarean section. This prevents the baby from coming into contact with infected material during its passage through the birth canal. In adults, meningitis and, more rarely, encephalitis may be seen as complications of genital herpes simplex infections.

7.3.8 Genital warts

Warts are benign tumours associated with papilloma virus infection. Genital warts, also known as condylomata accuminata, are associated with a number of strains, particularly types 6, 12, 16, 18 and 31. Warts develop from one to six months following infection and they occur on the penis, vulva or in the peri-anal region, depending upon the site of inoculation. Although at one time considered benign, if somewhat unsightly, the viruses that cause genital warts have been implicated as a cause of cervical carcinoma. This condition is particularly associated with warts caused by papilloma virus types 16 and 18. Lesions caused by papilloma viruses can be visualised as white plaques when 5% acetic acid is applied to the cervix. If these are detected, laser treatment can be used to prevent the possible development of cancer.

7.3.9   Pubic lice and scabies

There are three species of louse that infest humans: *Pediculus corporis*, the body louse, *Pediculus capitis*, the head louse, and *Phthirus pubis*, the pubic or 'crab' louse. Although it is mostly associated with the pubic region, the crab louse also colonises hair in the armpit, where it can cause considerable irritation. Lice lay eggs, known as 'nits' and these are firmly attached to pubic hair. It can be very difficult to remove nits from hair. Adult lice browse at the base of the hair. Carbaryl or malathion is applied to affected areas to treat an infestation.

Sarcoptes scabiei is a mite that may infest the local genitalia, causing the 'seven year itch'. These arthropods burrow under the skin to produce characteristic tracks. Patients with genital scabies typically have scabies lesions elsewhere on their body. Most often scabies affects areas between the fingers and in the toe webs. Although scabies is most often seen in people with poor standards of cleanliness, it may affect anyone. All sexual contacts should be treated to avoid 'ping-pong' infections, where partners pass the infection backwards and forwards between one another. Benzyl benzoate can be used to treat scabies.

7.4   What causes infections of the central nervous system?

The importance of the brain is illustrated by the fact that one medical definition of death is absence of brainstem function. The brain is highly protected. Anatomically it is provided with its own bone box, the skull. Physiologically it is also protected by the blood–brain barrier. This comprises the endothelial cells of the cerebral blood supply, astrocytes and the basal lamina of the brain. Together, these make a formidable barrier. Even so, infected lymphocytes and macrophages can cross the blood–brain barrier to introduce pathogens into the substance of the brain.

Because of its central importance, the brain makes enormous resource demands. To accommodate these, it must be well supplied with nutrients and its waste products must be removed. This also leaves the central nervous system vulnerable to infection since pathogens can exploit these transport systems to cause disease. Some pathogens can invade the brain by a spreading infection of the peripheral nervous system Infections of the central nervous system represent a major breach of the host defences and can range from the inapparent to the dramatic and rapidly fatal.

Infection of the membranes surrounding the brain, the meninges, is referred to as meningitis. When the infection affects the substance of the brain this is known as *encephalitis*. Brain abscesses are defined, localised lesions within the brain. The brain may also be subject to 'slow' infections that are generally progressive and fatal. Because of the importance of the brain and since infections of the central nervous system are potentially rapidly fatal, the rapid diagnosis of infections of the central nervous system is a high priority.

7.4.1   What causes meningitis?

Viruses cause most cases of meningitis and these may only be treated symptomatically. Complete recovery from virus meningitis is the rule. In contrast, bacterial meningitis is rare but it is a life-threatening condition. Fungi and protozoa can also cause meningitis but these are very rare. Whatever the cause, the clinical symptoms of meningitis are similar. Patients complain of a severe headache and have a fever. They suffer neck stiffness because of inflammation of the spinal cord and are **photophobic**, unable to tolerate bright light. Fast and accurate diagnosis of meningitis is essential because meningitis can be caused by a wide array of microbes and the outlook can vary from complete recovery to death, depending upon the causative agent. This permits rapid initiation of appropriate antimicrobial therapy, if appropriate.

Dating from a time when viruses were not recognised, virus meningitis is sometimes also known as aseptic meningitis since routine bacteriological culture of cerebrospinal fluid is negative. The most common causes of virus meningitis are enteroviruses, particularly echoviruses, coxsackieviruses and poliovirus, although meningitis is also a complication associated with mumps infection. Herpes simplex and other viruses can more rarely cause meningitis. Patients generally recover completely from meningitis caused by viruses. Treatment of virus meningitis is aimed at relieving the symptoms of the disease rather than trying to cure the infection.

Bacterial meningitis is always a severe and life-threatening infection requiring accurate diagnosis and prompt initiation of an appropriate antibacterial therapy. Other than in newborn babies, three pathogens are responsibly for the majority of cases of bacterial meningitis. These are *Neisseria meningitidis*, *Streptococcus pneumoniae* and *Haemophilus influenzae* of the Pittman type B, although with the introduction of the Hib vaccine, cases of *Haemophilus influenzae* meningitis have fallen dramatically. Between them, these three species account for 80% of the cases of bacterial meningitis.

Newborn babies are vulnerable to meningitis caused by Lancefield group B streptococci (a grouping based on cell wall antigens), coliform bacteria and *Listeria monocytogenes*. Meningitis caused by *Haemophilus influenzae* typically occurs in infants between the ages of six months and two years. Before babies are six months old, passively acquired maternal antibodies protect them from this bacterium and after two years children have acquired their own immunity to infection. The period between these two ages offers a window of opportunity for haemophilus infection. Older children and young adults are at risk of developing meningitis caused by *Neisseria meningitidis*, particularly if they live in overcrowded conditions. People of any age are potential victims of meningitis caused by *Streptococcus pneumoniae*. All the bacterial pathogens causing meningitis have a capsule that acts as a virulence factor.

In newborn babies, the diagnosis of meningitis can be very difficult. They do not show the typical and obvious clinical signs associated with meningitis. The symptoms they suffer may be very vague: 'failure to thrive' or being 'off their feeds'. Although babies may have a fever they may suffer **hypothermia** instead. They often suffer from diarrhoea at the same time.

Some women carry group B streptococci as part of their vaginal flora. These bacteria may infect babies during their passage through the birth canal. The risk of a baby becoming infected is increased if the infant is premature or if labour is protracted. Newborn babies infected with group B streptococci may go on to develop meningitis. *Listeria monocytogenes* is a Gram-positive bacillus that is widespread. It is sometimes found in large numbers in pâté and in soft cheeses. Pregnant women are discouraged from eating these foods because there is a risk that they will cause infection of the fetus. Newborn babies are also at risk from listeria infection. During the first week of life listeriosis generally presents as septicaemia, but thereafter babies infected with *Listeria monocytogenes* are likely to develop meningitis. Penicillin in combination with gentamicin is used to treat meningitis caused by group B streptococci and *Listeria monocytogenes*.

Although a pathogen of the respiratory tract, the Gram-negative coccobacillus *Haemophilus influenzae* can cause meningitis of insidious onset in babies unprotected by the Hib vaccine. It is associated with a greater risk of serious, permanent neurological damage than other bacterial meningitides. Complications include deafness, blindness, motor problems and learning difficulties. Treatment of meningitis caused by *Haemophilus influenzae* was usually with ampicillin but a significant minority of isolates produce a β-lactamase that can inactivate ampicillin. Chloramphenicol or an extended-spectrum cephalosporin such as cefotaxime can be used as alternatives. How long extended-spectrum cephalosporins will fulfil this role is uncertain. The gene that confers resistance to ampicillin in *Haemophilus influenzae* has, in other bacteria, mutated so that it also confers resistance to cephalosporins. It is surely only a matter of time before the gene in *Haemophilus influenzae* also mutates to broad-spectrum activity. The close contacts of a child with haemophilus meningitis are offered rifampicin to prevent spread of the infection.

About 20% of the population carry *Neisseria meningitidis* as part of the commensal flora of the throat. The bacterium can survive inside polymorphonuclear leukocytes. When people live in crowded conditions, such as in boarding schools or in military barracks the carriage rate is much higher, sometimes approaching 100%. The bacterium is spread by inhalation of infected droplets. It is perhaps no coincidence that following the rapid rise in their numbers, university students in the UK have recently become particularly vulnerable to meningococcal meningitis. This is a problem that is exacerbated when many people are crowded together in badly ventilated rooms such as lecture theatres. Very rarely, the healthy carrier status is disturbed and meningitis can result. The factors that cause conversion from carriage status to disease are not understood but the progress of the disease is extremely rapid. From the first signs that a patient is unwell to death may be only a few hours.

In about 80% of cases of meningococcal meningitis, patients also develop a rash. Initially this resembles purple bruising of the skin but later the lesions turn black as the damaged tissue dies. The rash does not disappear when pressed with a drinking glass and is evidence of meningococcal septicaemia. In about one third of cases patients also develop **endotoxic shock**. This may lead to **disseminated intravascular coagulation**, multiple organ failure and death. Upon **lumbar puncture** the cerebrospinal fluid may appear bloodstained if the patient is suffering from endotoxic shock. If it is not treated, meningococcal meningitis is invariably fatal. The mortality rate drops to 10% if appropriate chemotherapy is initiated promptly, with penicillin or an extended-spectrum cephalosporin being the treatments of choice. These may be combined with chloramphenicol: a drug that is associated with rare yet serious side effects. Close contacts are offered rifampicin as **prophylactic** treatment.

The antigenic structure of the meningococcal capsule is used to type these bacteria. There is a vaccine that is effective against both groups A and C meningococci. These strains are predominant in parts of Africa and travellers to these regions can be vaccinated to prevent them from acquiring meningococcal meningitis. In the UK, the predominant strains belong to groups B and C with group B meningococci occurring most frequently. The current vaccine is inactive against group B meningococci.

Encapsulated strains of *Streptococcus pneumoniae* can cause meningitis in people of any age but is more commonly seen in children under two years of age and in the elderly. Others at risk of developing pneumococcal meningitis include chronic alcoholics, people with sickle cell disease and those who have had their spleen removed. Pneumococcal meningitis may follow an attack of pneumonia or as a result of direct spread from an ear infection to the meninges. Streptococcal meningitis is associated with a mortality rate of between 20 and 30% even when treated appropriately. Some 15–20% of patients who survive suffer permanent neurological damage. There are vaccines that protect against pneumococcal disease but these are less effective in people who are at greatest risk of developing pneumococcal meningitis. The treatment of choice for pneumococcal meningitis is penicillin. There are, however, strains emerging that are tolerant to penicillin as a result of altered penicillin-binding proteins. Treatment of these strains is difficult.

Mycobacterium tuberculosis can occasionally cause meningitis that is of insidious onset. The slow development of symptoms can make diagnosis very difficult and this can delay the start of treatment. In turn, this increases the risk that the patient will die or will suffer severe complications because of the infection. Very rarely, meningitis caused by *Mycobacterium tuberculosis* can be of sudden onset. In such cases the clinical symptoms resemble sub-arachnoid haemorrhage, with the patient becoming disorientated and lapsing into unconsciousness. Almost always, patients with tuberculous meningitis will also have a focus of infection elsewhere. Most often the patient will have a lesion in the lung. Lesions may also occur in the spine, where damage to bone and the intervertebral discs can lead to nerve compression and, ultimately, paralysis. Therapy should include both isoniazid and rifampicin, both of which can cross the blood–brain barrier. Other antibiotics may be used, particularly if the causative strain is drug resistant.

Cryptococcus neoformans is an encapsulated yeast with round cells. It can be isolated in large numbers from pigeon droppings and is a rare cause of chronic meningitis that has an insidious onset, although the primary site of infection is in the lungs. Cryptococcal meningitis occurs sporadically throughout the world and is particularly associated with people suffering from full-blown

AIDS. Up to 20% of such patients develop the disease, although in Europe the figure is 3–5%. The course of the disease can run for months or years and it is nearly always fatal if untreated. AIDS patients tend to suffer fewer symptoms than do other victims. Amphotericin B and flucytosine are used to treat cryptococcal meningitis. Patients with AIDS tend to suffer relapses and may also react badly to the normal treatment. Fluconazole may be used in cases of treatment failure and it is also used in AIDS patients as maintenance therapy to prevent relapses.

Another fungus, *Coccidioides immitis*, causes a chronic meningitis that is treated with amphotericin B. This fungus is only a problem on the American continent. Amoebic meningitis, caused by *Naegleria fowleri*, has been discussed in Section 3.1.6.

How is meningitis diagnosed in the diagnostic microbiology laboratory?

If a patient is suspected of having meningitis the most important procedure is to perform a lumbar puncture to obtain a specimen of cerebrospinal fluid. Biochemical and microbiological tests can then yield valuable information very rapidly (Table 7.3). Lumbar puncture is a procedure that carries a significant risk. If a patient has a brain abscess rather than meningitis, then there is a raised fluid pressure within the skull. If this pressure is suddenly released, for example when a lumbar puncture needle is introduced into the spine, then brain tissue is forced downwards out of the skull. Brain matter is pushed through the foramen magnum: the hole through which the spinal cord passes. This causes a fatal 'cerebellar cone'. If there is a risk that the patient has a brain abscess then a brain scan should be carried out before the lumbar puncture to exclude this possibility. Along with a cerebrospinal fluid sample, blood cultures should be taken since the bacterial pathogens that cause meningitis most frequently case septicaemia as well. Blood cultures increase the likelihood of isolating the causative agent.

Important information can be obtained by noting the appearance of the cerebrospinal fluid sample providing the lumbar puncture is a 'clean take' and not contaminated with blood as a result of poor sampling. Normally cerebrospinal fluid is clear and looks like water. If the sample is slightly opalescent, then virus meningitis or a chronic meningitis such as that caused by *Mycobacterium tuberculosis* or *Cryptococcus neoformans* may be suspected. There is not such a marked cellular response to these infections compared with acute bacterial meningitis. In such cases the cells in the cerebrospinal fluid make the sample turbid (cloudy). The cerebrospinal fluid samples from patients with tuberculous meningitis may contain a 'spider's web clot'. Blood-stained cerebrospinal fluid

Table 7.3 *Summary of the examination of cerebrospinal fluid samples from patients with meningitis*

Causative agent	Appearance	Protein	Glucose	Microscopy Cells	Microscopy Microbes
Neisseria meningitidis	Turbid	High	Low	Polymorpho-nuclear leukocytes	Intracellular Gram-negative diplococci
Streptococcus pneumoniae	Turbid	High	Low	Polymorpho-nuclear leukocytes	Lanceolate Gram-positive diplococci
Haemophilus influenzae	Turbid	High	Low	Polymorpho-nuclear leukocytes	Gram-negative cocco-bacilli
Mycobacterium tuberculosis	Opalescent	High	Slightly low	Lymphocytes	Acid alcohol-fast bacilli
Cryptococcus neoformans	Opalescent	Raised	Slightly low	Lymphocytes	Yeast cells surrounded by a clear halo in India ink preparations
Viruses	Opalescent	Raised	Normal	Lymphocytes	Not usually performed

samples may indicate a bad specimen but also suggest sub-arachnoid haemorrhage.

Biochemical tests are performed to determine the level of protein and relative quantities of glucose. For this test it is useful to establish the concentration of glucose in the blood at the time the cerebrospinal fluid sample is taken, since this can fluctuate throughout the day. With virus meningitis and chronic meningitides, protein levels are raised but the level of glucose is normal or only marginally depressed. In acute bacterial meningitis, protein levels are significantly raised above the normal value of 14–45 mg/dl and glucose levels are severely depleted: generally being less than 40% of the serum glucose concentration. It was at one time thought that this was caused by bacterial metabolism, but alterations in brain chemistry because of infection also contributes significantly to the depletion of glucose in cerebrospinal fluid.

Microscopic observation of cerebrospinal fluid allows the cells present in a sample to be identified. In acute bacterial meningitis, polymorphonuclear leukocytes predominate, but in virus meningitis and meningitis caused by *Mycobacterium tuberculosis* or *Cryptococcus neoformans* lymphocytes are present in excess. Since *Cryptococcus neoformans* is an encapsulated yeast, it can easily be visualised by adding India ink to the infected cerebrospinal fluid. The capsule keeps out the particles of ink and the yeast cells can easily be seen: each surrounded by a clear halo.

Examination of a Gram-stained centrifuged deposit from cerebrospinal fluid may lead to a rapid diagnosis but care must be taken with this test to ensure appropriate decolorisation. If not then the cocco-bacilli of *Haemophilus influenzae* may be confused with *Streptococcus pneumoniae*. If tuberculous meningitis is suspected, then a Ziehl Neelsen or the fluorescent Auramine–Rhodamine stain for acid alcohol-fast bacilli should be performed. Electron microscopy is rarely carried out in cases of virus meningitis since it is an expensive and technically demanding technique, the results of which will not affect patient management.

Bacterial pathogens implicated in meningitis may be difficult to grow in artificial culture. Specimens are inoculated onto heated blood agar plates and incubated under 5–10% carbon dioxide. cerebrospinal fluid samples may, however, be culture negative even in a frankly infected patient. This is a particular problem if patients have been given an antibiotic before the cerebrospinal fluid sample is taken. Saving a life is, on balance, more important than obtaining a good clinical specimen. Time is not on the side of the clinician when dealing with acute bacterial meningitis. Antigen detection kits may be used to identify the causative agent of meningitis rapidly from a cerebrospinal fluid sample but these are relatively insensitive and are not often of much direct use in patient management. Although *Mycobacterium tuberculosis* is very slow growing in artificial culture, taking up to eight weeks to culture, PCR-based technology can now identify cerebrospinal fluid samples infected with this bacterium in less than one day.

7.4.2   What causes encephalitis?

Infection of the substance of the brain is referred to as encephalitis. Symptoms of this infection include changes in the person's behaviour, seizures and an altered level of consciousness. Patients often also complain of nausea and this may lead to vomiting. Viruses are the most common cause of this condition. By using isotopic tracing with technetium-99, brain scans can be used to show areas of inflammation. There should be a correlation between the scan results and the neurological abnormalities that the patient suffers, depending upon the areas of the brain that are affected.

Herpes simplex virus causes most cases of encephalitis. In newborn babies, herpes simplex virus can cause an overwhelming and generalised infection that also affects the brain. Infection is acquired during vaginal delivery when the mother has a genital infection. Encephalitis is rare in adults and is usually caused by a reactivation of latent infection of the trigeminal ganglion. In adults, the symptoms reflect the area of the brain that is affected. If herpes encephalitis is left untreated it has a 70% mortality rate. Aciclovir therapy has reduced this figure dramatically. Two other herpesviruses, varicella zoster and cytomegalovirus are rare causes of encephalitis. Zoster (shingles) affecting the optic nerve can rarely lead to encephalitis. Cytomegalovirus can cause encephalitis in AIDS patients as latent infections are reactivated. Rarely, cytomegalovirus can infect a fetus in the womb. When this happens the virus can cause gross destruction of the fetal brain. This leads to the birth of severely brain-damaged infants.

Over thirty flaviviruses cause an arthropod-borne infection of the brain that leads to encephalitis. The most dramatic is the mosquito-borne Japanese encephalitis virus. This causes epidemics across Asia, even though a vaccine for this virus exists. Other arthropod-associated encephalitides tend to be named after the geographical location in which they are first reported: St Louis encephalitis virus, Murray Valley encephalitis virus. Alternatively, they are named after the vector, as with tick-borne encephalitis virus

Although mumps virus is typically associated with infection of the salivary glands, it may also cause both meningitis and encephalitis without salivary gland involvement. Patients with full-blown AIDS often develop a form of pre-senile dementia. The brain substance shrinks as the brain ventricles enlarge. This is a consequence of sub-acute encephalitis caused by HIV. Clinically it can be difficult to differentiate this from progressive multifocal leukoencephalopathy, another brain infection seen in AIDS patients.

Measles and rubella are both typically self-limiting diseases characterised by a generalised rash over the entire body. Both infections are, however, associated with a rare complication: **sub-acute sclerosing panencephalitis (SSPE)**. This is a slow and progressive disease that is ultimately fatal. Its first symptoms occur about a decade after the initial virus infection. Virus RNA persists in the brain and can be detected either from brain biopsy material or at *post mortem* examination using PCR-based technology. Measles virus genomes have even been detected in museum specimens preserved for many years, embedded in paraffin wax and preserved for histological rather than molecular biological examination.

Bacteria may cause encephalitis very rarely. The organisms associated with this condition include *Legionella pneumophila*, *Borrelia burgdorferi* and *Treponema pallidum*. All can cause a primary encephalitis. The symptoms of tertiary neurosyphilis are immunological in origin rather than a direct consequence of infection. Other rare causes of encephalitis include the yeast *Cryptococcus neoformans* and the malarial parasite *Plasmodium falciparum*, as well as trypanosomes, the cause of sleeping sickness.

The protozoan parasite *Toxoplasma gondii* generally causes a mild or subclinical infection. If, however, a pregnant woman is infected, the protozoa may cross the placenta to infect her fetus. If this happens, the fetus may suffer a generalised infection, including an encephalitis. Clinical symptoms of toxoplasmosis include epilepsy and blindness. Since kittens and puppies excrete these protozoa, pregnant women are advised not to change cat litter trays, or are at least advised to wear gloves when so doing and to wash their hands thoroughly afterwards.

7.4.3   What is rabies?

Rabies is an invariably fatal infection of humans acquired through a bite of an infected animal. Worldwide there are estimated to be 75 000 cases of human rabies each year. Its island status and its strict quarantine laws protect the UK. Quarantine is the period of isolation imposed on animals to ensure that they do not develop symptoms of disease: six months in the case of rabies. Its name is derived from the Italian, meaning forty days, since during the Renaissance this was the period of isolation usually imposed on individuals suspected of harbouring plague. Since the eradication of rabies from the UK in 1906 only very occasional cases have occurred here. Although rabies is fatal for humans, the virus causes a mild and even sub-clinical disease in a wide variety of 'lower' mammals, including dogs, wolves, foxes, bats, skunks and jackals. The last case of animal rabies in the UK was in 1996, when a bat flew across the English Channel to land near Newhaven in Sussex. Such importations are very rare, even if they are reminiscent of Dracula's fictional landing at Whitby. Customs officers are, however, constantly vigilant in preventing the importation of smuggled pet animals. New UK legislation is currently being considered to lift the quarantine requirement, relying instead upon animal vaccination to provide protection.

If a human is bitten by a rabid animal, then the wound should be thoroughly cleaned immediately and the victim given rabies-specific immunoglobulin to help to minimise the risk of developing rabies. Following the bite from a rabid animal, the rabies virus is inoculated into a peripheral nerve. Over the next few months the virus travels up the infected nerve until it reaches the central nervous system. The initial symptoms of rabies include a sore throat, fever and headache. Victims then develop muscle spasm and convulsions. Patients become increasingly disoriented before death. Nerve damage affects the swallowing reflex. This may be why rabies was once known as hydro-phobia: fear of water. The mere sight of water may induce terror in a patient.

The virus may be detected using a fluorescent antibody test. Corneal scrap-ings can be used for this, as can brain biopsy material. Examination of the brain at *post mortem* reveals that the cells of the hippocampus contain characteristic inclusion bodies known as **negri bodies**.

7.4.4 What is progressive multifocal leukoencephalopathy?

The papovavirus JC typically causes a sub-clinical infection sometime during childhood and the majority of people across the world have been infected with this virus. Its DNA can be detected in healthy human tissue, in which the virus causes a persistent sub-clinical infection. JC virus can, however, reacti-vate in patients with severe immunodeficiency. Reactivation of infection to cause progressive multifocal leukoencephalopathy is typically seen in patients with lymphoma or who have developed full-blown AIDS. About 4–5% of AIDS sufferers will develop progressive multifocal luekoencephalopathy, of whom 80% will die within a year. The pathology of the disease is character-ised by the development of a number of white plaques. These occur through-out the brain and represent areas where the myelin that normally surrounds nerve cells has become destroyed. Accompanying this is a progressive loss of nerve function and increasing neurological disease: weakness, impaired memory, apathy and cognitive disorders. This leads relentlessly to death.

7.4.5 What are poliomyelitis and chronic fatigue syndrome?

In the vast majority of people infected with poliovirus, the worst symptoms that they experience are a mild sore throat and, perhaps, diarrhoea. In about 1% of people infected with this virus the symptoms are more severe. Patients develop a mild, self-limiting form of meningitis. In very few individuals, however, poliovirus causes paralytic poliomyelitis and associated muscle wasting. This results from infection and damage of the motor nerves. Ironically, the risk of developing paralytic poliomyelitis is greatest in active young people. There is no way of predicting which muscles will be affected. If the respiratory muscles are damaged, patients cannot breathe. Before the successful introduction of polio vaccination, patients with paralytic poliomyelitis affecting the respiratory muscles were condemned to a life in an 'iron lung'. In such a device, the patient's body is enclosed in an airtight chamber, leaving just the head exposed. To mimic breathing, the air pressure within the chamber is reduced, forcing the chest to expand, dragging air into the lungs. The pressure is then normalised and the chest collapses, forcing air in the lungs to be exhaled. Breathing with an iron lung is a very painful experience. Where leg muscles have been affected, victims have been confined to a wheelchair for life.

The first vaccine developed against the poliovirus was the Salk vaccine, which used killed virus particles. The Sabin vaccine, a live, attenuated vaccine, quickly superseded this. When this vaccine was first used in the late 1950s children loved it, since it was administered on a sugar lump and not through a needle. Use of this vaccine has now eradicated poliomyelitis from the North American continent and much of the rest of the world. Indeed, it is likely that poliomyelitis will be the second infectious disease to be eradicated from the planet: the first being smallpox. At present the World Health Organization are organising a series of 'immunisation days' when many millions of individuals throughout the world are vaccinated against poliomyelitis on the same day. When delivering the vaccine, care must be taken to vaccinate all vulnerable household contacts. There have been two incidents where unprotected close family contacts of babies given the Sabin vaccine have developed paralytic poliomyelitis as the vaccine strain has undergone the double mutation to revert to virulence. In the early days of polio vaccination, the cell lines used to grow the virus were contaminated with a monkey virus, SV40. This virus is a papovavirus related to JC and BK viruses. Live SV40 virus was copurified with the vaccine strain of poliovirus. Although SV40 was administered to innumerable vaccination candidates across the world in this way, there was not a single reported case of human SV40 infection.

There is evidence suggesting that at least some cases of chronic fatigue syndrome may be caused by persisting enterovirus infection, perpetuated by the production of defective virus particles. This condition is also known as post-viral fatigue syndrome, myalgic encephalomyelitis and by many other names. Although a number of viruses have been linked with this condition, characterised by overwhelming muscle fatigability and pain, the most common viruses to be associated belong to the coxsackievirus family. These viruses are close relatives of the poliovirus. In the 1940s and early 1950s when paralytic poliomyelitis was epidemic, cases of chronic fatigue syndrome were known as 'abortive poliomyelitis'. This was because of the similarity of the symptoms of the two diseases, although chronic fatigue syndrome is not associated with muscle wasting.

7.4.6 What are transmissible spongiform encephalopathies?

The transmissible spongiform encephalopathies (TSEs) have been in the news throughout the 1990s through the 'mad cow disease' tragedy. Mad cow disease is more properly known as bovine spongiform encephalopathy, or BSE for short, and it first appeared in 1987. The name derives from the characteristic sponge-like holes that are seen upon microscopic examination of the brain at *post mortem* examination. Cattle with BSE become nervous and difficult to handle. They also become unsteady on their feet. The pathology of this disease is similar to scrapie, a disease of sheep that has been recognised for at least 200 years. Sheep with scrapie appear irritable and have an insatiable desire to scratch themselves against fences, walls, gates, etc. – hence the name. At *post mortem* examination, brains from sheep with scrapie also have a sponge-like appearance, although the lesions are distinct from those of BSE. Other animals are also affected by transmissible spongiform encephalopathies. These include transmissible mink encephalopathy seen in farmed mink and chronic wasting disease of musk deer and elk.

These diseases are thought to be caused by **prions**. These have been defined as proteinaceous infectious particles that are resistant to procedures used to degrade nucleic acids. The precise nature of prions is the subject of considerable debate and there are still eminent scientists working in this area that do not believe in the prion theory. It is now becoming accepted that a prion is the modified form of a normal cellular protein PrPc. The prion form, PrPsc, is relatively resistant to protease digestion and accumulates in fibrils in infected brains. It is also thought that introduction of PrPsc into normal tissue causes conversion of the normal PrPc protein into the pathological PrPsc form, by a mechanism that has yet to be explained.

It is not certain how BSE arose but it is thought that the first cattle to suffer from BSE were fed material that contained the brains from scrapie-affected sheep. Furthermore, the rendering process used in food production had not inactivated the infectious agent. In this way the infectious agent is thought to have crossed the species barrier. During the early 1990s, cats and a variety of zoo animals developed spongiform encephalopathies after being fed on meat from BSE-affected cattle.

There are several human diseases with a pathology similar to scrapie and BSE. These include Creutzfeldt–Jakob disease, Gerstmann–Straussler–Scheinker syndrome, fatal familial insomnia and kuru. Symptoms of these diseases include loss of motor control, dementia and a wasting paralysis progressing over months to an inevitable death. As the name fatal familial insomnia implies, there is a genetic disposition to at least some of these diseases. Other human spongiform encephalopathies appear to behave like infectious diseases. It has been estimated that at the time of death, about one person in ten thousand has been infected with the agent that causes Creutzfeldt–Jakob disease. It has long been recognised that Creutzfeldt–Jakob disease can be passed on from one individual to another by using inadequately sterilised neurosurgical instruments or through corneal grafts. It was the practice to purify certain hormones from pooled pituitary glands collected during *post mortem* examinations. Use of human growth hormone derived from this source has been linked with the subsequent development of Creutzfeldt–Jakob disease. Kuru, meaning 'trembling with fear', is associated with cannibalistic funeral customs of New Guinea Highland tribes.

These human spongiform diseases have a very long incubation period. Symptoms may take between 10 and 30 years to appear. There is currently a raging debate concerning a new form of Creutzfeldt–Jakob disease. The new variant form affects young people; some are only in their teen years. Classically, Creutzfeldt–Jakob disease is an illness that affects ageing people. Furthermore, the lesions in the brains of people dying from new variant Creutzfeldt–Jakob disease are distinct from the classic form of the disease. They resemble the lesions seen in cattle who develop BSE. Although people were at fist dismissive of a link between BSE and Creutzfeldt–Jakob disease, it is now accepted that the new variant Creutzfeldt–Jakob disease is most likely acquired by eating contaminated beef products. Beefburgers and sausages pose a special risk because they contain unspecified 'meat' and at one time this included brain tissue. After an early BSE scare, the nervous tissue of cattle was removed from the human food chain.

This raises the spectre of an epidemic of new variant Creutzfeldt–Jakob disease. Very many people were consuming contaminated beef products in the early years following the first appearance of Creutzfeldt–Jakob disease. Will this happen? No one can currently predict with certainty. There are too many factors that can influence the situation. Certainly many people will have been exposed to contaminated beef products. How infectious is this disease for humans? What is the size of inoculum required to cause disease in humans who eat contaminated food? How high is the cow–human species barrier to infection? We will not know the answers to these questions until much more research has been carried out. This may take many years and even then some questions can only be answered by examining circumstantial evidence. It is hardly ethical to conduct feeding experiments to establish the infective dose, for example.

7.4.7 What causes brain abscesses?

Brain abscesses are defined lesions within the substance of the brain. Most often they occur as complications following surgery or accidental trauma but may result from infections extending from other close anatomical sites. Examples are abscesses associated with ear infections or sinusitis. Brain abscesses cause a rise in fluid pressure in the skull and should be ruled out before a patient undergoes a lumbar puncture procedure. If not, then patients with brain abscesses are likely to die as brain matter is forced out of the skull when the intra-cranial pressure is released.

Brain abscesses are rare and are most often caused by more than one organism. They are often associated with anaerobic bacteria. Brain abscesses are typically caused by actinomycete bacteria and members of the genus *Nocardia*. Brain abscesses associated with dental infections are commonly caused by members of the genus *Bacteroides* together with microaerophilic streptococci such as *Streptococcus anginosus*. Even rarer are brain abscesses caused by bacteria including *Mycobacterium tuberculosis, Treponema pallidum* and *Borrelia burgdorferi*. Fungi including *Cryptococcus neoformans,* members of the genus *Aspergillus, Coccidioides immitis, Candida albicans, Histoplasma capsulatum* and *Blastomyces dermatitidis* have all very occasionally been implicated as causes of brain abscesses. Treatment of brain abscesses involves surgical drainage of the lesion as well as an extended course of the appropriate antimicrobial agent.

7.4.8 What is tetanus and how is it related to botulism?

There are two important bacterial intoxications that affect the central nervous system, tetanus and botulism. Anaerobic, spore-forming bacteria belonging to the genus *Clostridium* cause both diseases. *Clostridium tetani* causes tetanus and *Clostridium botulinum* is responsible for botulism. Both cause paralysis. In the case of botulism, muscles relax whereas in tetanus muscles are forced into spasm. Botulism is discussed in Chapter 5.

Tetanus is caused when spores of *Clostridium tetani* are inoculated into wounds. Wounds can range from massive injuries contracted in battles to minor scratches acquired during gardening. In some cases, the injury may be so slight as to be inapparent. Approximately one million babies die every year because they acquire tetanus from contamination of their umbilical stump following birth. Tetanus toxin attaches to the nerve synapse of motor neurones where it blocks the release of glycine, a factor important for muscle relaxation. This causes the muscles to remain in a permanent state of contraction. The intensity of muscle contraction can cause tissue damage and this leads to death. The mortality associated with tetanus is about 50%, although this has been considerably reduced in developed countries by immunisation with tetanus toxoid.

7.5   What causes infections of the circulatory system?

The circulatory system has evolved to act as a transport system delivering nutrients and removing waste products from around the body. As such, it can potentially act as a vector to spread infection as well. To counter this threat, we have evolved an elaborate system of defences that make blood an inhospitable environment for microbes. Humoral defences, including immunoglobulins, complement, cytokines and so on, protect us. If these fail to stop the invading microbes then we can rely on our cellular defences, such as phagocytic leukocytes. Our defences are probed on a regular basis, each time we brush our teeth vigorously, for example, or even chew on a sticky toffee. Small numbers of microbes regularly gain access to our bloodstream through such activities. Typically, our innate defences rapidly remove them. Occasionally, however, an overt infection is established. This leads to a life-threatening condition.

To overcome our innate defences, the bacteria that cause infections in our circulatory systems must overcome considerable challenges. One answer is to establish a very large initial inoculum so that the sheer numbers of invading organisms overwhelms our defences. This does not happen very often. More likely, bacteria exploit weaknesses in our defences. Often these weaknesses occur because of the introduction of urinary or vascular catheters, surgical instruments and other medical devices such as artificial ventilators. These all breach our natural anatomical defences and provide easy access to our circulation for microbes. For similar reasons, burn victims are vulnerable to septicaemia because of the damage to skin, an excellent natural defence against infection.

7.5.1   A problem with terminology

The literature describing infections of the circulatory system is plagued by a confused terminology. Words tend to be used as if they were interchangeable and this may lead to considerable confusion. **Bacteraemia** is the presence of bacteria in the bloodstream. This is a common occurrence and is usually inapparent. Some authors, however, use this term to describe occasions when bacteria in the bloodstream cause a clinically significant response. Others prefer to use the term septicaemia to describe the clinically apparent presence of microbes of any description in the blood. In this book, the term bacteraemia is used to refer to the presence of bacteria in the bloodstream and septicaemia is used to describe the *clinically significant response* to the presence of microbes in the blood. The vague term 'sepsis' can also cause confusion. Some authors restrict the definition of 'sepsis' to conditions associated with septicaemia while others use the word to describe any infectious process.

The presence of viruses in the bloodstream is referred to as viraemia. It is in this manner that viruses spread throughout the body and thus they can attack their target organ. In severely compromised people, fungi can enter the bloodstream in significant numbers. This is referred to as **fungaemia**. Some parasites also live inside the bloodstream of humans for at least part of their lifecycle. These include the plasmodium family, trypanosomes and *Leishmania* species. When parasites such as these are found in the blood, the condition is described as **parasitaemia**.

7.5.2   What is plague?

One of the greatest influences a microbe has had on the history of the Western civilisation has surely been wrought by the plague, caused by the Gram-negative bacillus *Yersinia pestis*. Mycobacteria may have influenced fashion for a short period when high collars were worn to hide lesions of scrofula. Syphilitic lesions were also disguised with make up during the seventeenth century. Yet the plague held Europe in a devastating grip during the middle ages and well into the seventeenth century. The Book of Common Prayer of 1662, still used today by the Church of England, specifically includes a 'Thanksgiving for Deliverance from the Plague and Other Common Sicknesses'. This is an honour not afforded to any other infection.

Yersinia pestis infection is endemic in at least 200 species of rodent across the world. They suffer from sylvatic plague. The name is an indication that these animals live in the countryside: *sylvestris* is Latin for woodlands. These rodents act as a reservoir of infection of the black rat, *Rattus rattus*. Bites from the fleas that live on these rats transmit *Yersinia pestis* from infected animals to humans, particularly in conditions of poor sanitation. In Elizabethan England, for example, during episodes of plague measures were taken '. . . to warn inhabitants to keep channels against their houses free from filth'. Furthermore, they were advised '. . . not to make dunghills out of the stables in the streets'.

Ticks, lice and bed bugs have been shown also to act as vectors of plague, but spread other than by fleas is rare. Fleas gorge on the blood of an infected rat. They thus acquire bacteria circulating in the bloodstream of the rat. Bacteria multiply within the flea gut. If rats are not immediately available, humans will provide an alternative supply of blood. The fleas can then inoculate the human victim during a subsequent feeding session.

The most common form of plague is bubonic plague. The site of entry may develop a pustule or carbuncle but this is not common. The infection localises in the lymph nodes causing them to swell. Subsequently, bacteria spread around the body. Patients suffer a high fever and prostration: consequences of septicaemia. This is accompanied by a rash as blood leaks from damaged blood vessels in the skin. The lymph nodes turn black as the surrounding tissue becomes affected, first by an area of haemorrhage, then, as the oxygen supply is diminished the tissue dies and blackens. The affected lymph nodes are referred to as **buboes**. The blackening of affected tissue gives plague its alternative name: the Black Death. A much less common form of plague is the pneumonic form, where patients acquire infection through inhaling infected droplets in the air breathed out by a plague victim. This primarily affects the respiratory tract and buboes develop on the tonsils.

The nursery rhyme 'A ring a ring of roses' is a chilling allusion to the plague . . .

> '*A ring a ring of roses,*
> *A pocket full of posies,*
> *Atishoo, atishoo,*
> *We all fall down.'*

The ring of roses refers to the rash that affects victims. Plague was quickly associated with insanitary conditions, but for many years it was thought to spread by a 'miasma' or noxious cloud. People believed that the harmful effects of a 'miasma' could be prevented using nosegays: bunches of flowers This is the pocket full of posies of the rhyme. Indeed English Judges still carry nosegays of freshly cut flowers at the Old Bailey and on ceremonial occasions, even though plague has long since departed from the UK. 'Atishoo, atishoo' refers to the sneezing bouts that plague victims suffer. 'We all fall down' shows that victims of the plague die. Although no longer pandemic, plague still occurs in areas of Asia, Africa and South America.

7.5.3 What causes septicaemia?

As medical practices change so do the causes of septicaemia. Until the middle of the twentieth century, septicaemia was most commonly associated with *Staphylococcus aureus* and *Streptococcus pyogenes*. These bacteria are sometimes referred to as **pyogenic cocci**, since they induce pus formation: *puon* is the Greek word for pus. Following the Second World War, and because of changes in medical practices, Gram-negative bacteria became the predominant cause of septicaemia. Most important were members of the Enterobacteriaceae, especially *Escherichia coli*. As medical science continues to evolve there is at present a re-emergence of Gram-positive bacteria as a significant cause of septicaemia.

People who suffer immunosuppression are particularly vulnerable to septicaemia. This may be the result of a disease process such as cancer, particularly leukaemia. Alternatively, immunosuppression may be deliberate as in the case of organ-transplant recipients who have immunosuppressant therapy to prevent rejection of the transplanted organ. Treatment of the cancer increases further the risk of a patient developing septicaemia. Chemotherapy reduces the number of leukocytes in the patient's blood. This leaves the patient particularly vulnerable to infection. Septicaemic episodes in such cases are often caused by bacteria that are common in hospital environments, including *Pseudomonas aeruginosa*, *Stenotrophomonas maltophilia* and acinetobacters. These all pose serious problems because they are typically resistant to the antibiotics commonly used to treat septicaemia.

Newborn babies are particularly vulnerable to septicaemia caused by Gram-positive bacteria. Particularly problematic are streptococci of the Lancefield group B and the non-sporing bacillus *Listeria monocytogenes*. Although septicaemia is a life-threatening condition, its diagnosis in the newborns can be difficult. This is because, as with meningitis, babies do not show typical clinical signs of disease.

7.5.4 What are the symptoms and consequences of septicaemia?

The symptoms of patients suffering from Gram-negative septicaemia are those associated with endotoxin. This is the lipopolysaccharide (LPS) that forms the outer layer of the outer membrane of Gram-negative bacteria. Endotoxin activity is associated with the lipid A part of the lipopolysaccharide molecule. It causes dilatation of the blood vessels. This allows blood to leak into the surrounding tissues. Patients feel warm because there is an increased blood flow at the surface of the body. As blood leaks from blood vessels, the patient's blood pressure becomes abnormally low. To compensate for this the heart rate increases dramatically. These effects are sometimes referred to as 'warm shock'. Patients look hot and flushed.

As the disease progresses, it becomes necessary to preserve the blood supply to the vital organs. The circulation to the skin becomes restricted as the body attempts to ration its resources. The skin then becomes cold to the touch and the patient appears very pale. This phase may be referred to as 'cold shock'. If the disease progresses further then damage to major organs occurs. Particularly vulnerable are the kidneys and lungs. To complicate matters, the lungs can become filled with tissue fluid, leading to a condition described as adult respiratory distress syndrome (ARDS). The fluid in the lungs prevents blood from being adequately oxygenated and so the downward spiral continues.

In about 10% of cases of Gram-negative septicaemia, patients develop disseminated intravascular coagulation as multiple blood clots form throughout the circulatory system. These lead to multiple organ failure as the blood fails to penetrate tissues. This deprives organs of oxygen and a means of disposing of metabolic waste. This condition is nearly always fatal. The symptoms of Gram-positive septicaemia cannot be explained by endotoxin.

The treatment of septicaemia does not rely on giving appropriate antimicrobial chemotherapy alone. Supportive measures to relieve the clinical symptoms are required. These may include artificial ventilation to assist patients with ARDS or dialysis in cases of renal failure, for example. Often, septicaemia results from bacteria spilling out from a focus of infection somewhere in the body. If septicaemia is to be controlled successfully, then the focus of infection must also be treated adequately.

Since septicaemia is a life-threatening condition potentially caused by a wide variety of microbes, the initial therapy has to be 'blind'. Patients are often given an aminoglycoside, together with a broad-spectrum β-lactam antibiotic. These antibiotic families have a **synergistic** effect (Fig. 7.3). This is where the combined activity of the drugs is greater than the sum of the activity of each drug when used alone. If the patient has an underlying problem involving the gut, metronidazole will also be included to cover the possibility of an anaerobic infection. Once the causative agent has been isolated in the laboratory, its antimicrobial susceptibility is determined and the antimicrobial therapy may

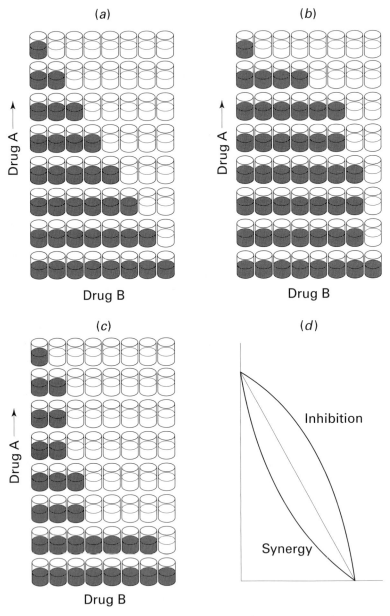

Fig. 7.3. The interaction between antibiotics. This can be tested in a 'checkerboard' experiment. Serial dilutions of two antibiotics are prepared in broth in the wells of a microtitre plate. The wells are then inoculated with a standard number of test bacteria and incubated appropriately. (a) If the antibiotics do not interact, a diagonal line will be seen regularly across the plate. (b) Some combinations of antibiotics interfere with each other as seen by growth at higher concentrations of the drugs than seen with antibiotics that do not interact. (c) Other combinations are synergistic and the combined effects of low concentrations of drugs prevent growth. (d) This can be represented graphically.

be adjusted as appropriate. Antibiotic therapy may have to be prolonged, depending upon the nature of the underlying condition.

7.5.5 How is septicaemia diagnosed in the diagnostic microbiology laboratory?

The laboratory diagnosis of septicaemia depends upon isolating the causative microbe in a blood culture. Great care must be taken with sample collection because the diagnosis of septicaemia has potentially grave consequences. There are a number of potential problems with sample collection. Blood must be extracted from the patient. This requires puncturing the skin and withdrawing the sample from a vein. The skin is not sterile and this allows an important opportunity for specimen contamination.

Because of its role in nutrient and waste transport, blood is protected by a rich array of antimicrobial defences. These could severely impair the chances of culturing microbes from blood samples. To overcome the effects of these inhibitors, blood samples are diluted. Typically, a 10 ml sample of blood is diluted in 100 ml of specialised broth. This culture is then appropriately incubated and observed. Typically, blood cultures are observed regularly over a one week period, but if a patient is suffering from fever of unknown origin or from endocarditis, incubation may be prolonged for up to one month. It is unwise to rely on the results obtained from a single culture. This may fail to grow or may yield a contaminant. In either case, the result is misleading. Most blood culture systems exploit more than one type of broth. A very wide array of microbes can be isolated from blood cultures and different broths are used to support the growth of different groups of microbes. For example, the broths used to isolate anaerobes have reducing agents added to them to provide a suitable environment in which to grow bacteria that cannot tolerate oxygen.

It is recommended that in the investigation of septicaemia, and of endocarditis, three separate blood cultures are examined. Ideally these should be taken at different times over a 24-hour period and the samples should be withdrawn from different sites. This maximises the chances of isolating pathogens and of identifying commensal microbes that are present in single cultures as contaminants. If all three samples grow the same coagulase-negative staphylococcus and these have been withdrawn from different sites this is a clear indication that the bacterium is a pathogen present in the bloodstream. If only one sample out of three yielded a coagulase-negative staphylococcus then this is much more likely to be the result of contamination of the sample. Similarly, if three different strains of coagulase-negative staphylococci were isolated in each of the three samples, then this is likely to represent contamination rather than polymicrobial infection. This is because coagulase-negative staphylococci are common skin commensals that easily contaminate blood samples. With severely ill patients the ideal of collecting three well-spaced samples may not always be possible. The needs of the patient must be considered paramount.

Septicaemia is often transient in nature. There is a constant fluctuation in the number of circulating microbes as they multiply and as the patient's defences attempt to eliminate the invaders. Consequently, the timing of blood collection is very important. Samples are best taken when a patient's temperature has just 'spiked'. The increase in body temperature indicates a response to the increase in number of circulating microbes.

Specimen collection requires careful attention. The skin at the site from which the sample is to be taken must be thoroughly cleaned with an alcohol-based antiseptic and it must then be allowed to dry adequately. Once the sample has been withdrawn it must be placed directly into the culture bottle, observing good microbiological practice. It is important that the cover and cap on the bottle are also sterilised. Often, other tests will need to be carried out on blood samples at a time when blood cultures are collected. An easy mistake that inexperienced blood samplers make is to distribute blood into other bottles before inoculating the blood culture. If this happens, there is a risk that the blood sample may become contaminated. This is a particular problem for blood that has come into contact with citrate solution, used as an anticoagulant to stop the blood from clotting. Citrate solutions often harbour Gram-negative bacilli such as *Serratia marcescens, Enterobacter* species, *Citrobacter* species and *Klebsiella* species. These can contaminate the sample and can yield apparently positive cultures. This phenomenon is sometimes referred to as **pseudosepticaemia** since the patient is not suffering from the infection indicated by the test result. This is a particular problem since these bacteria are common causes of infection in hospitalised patients.

There are several techniques used to detect microbial growth from a blood culture. The least satisfactory method is to take a loopful of the broth and to sub-culture this on a regular basis, since this introduces further chances to contaminate the blood culture. To overcome this problem, some laboratories use a biphasic medium. A suitable agar-based medium is poured into the culture bottle and is allowed to set as a slope up the side of the vessel. Broth is then introduced into the bottle, so that part of the solid medium is exposed above the surface of the liquid. Once the blood sample has been inoculated the vessel is a closed system. The bottle is then tilted and set upright again so that the broth washes over the solid medium. In this way the exposed solid medium can become inoculated with microbes from the sample. This is repeated as necessary and any growth is observed. Alternatively, microbial growth may be detected indirectly. This may be by observing the release of radiolabelled carbon dioxide derived from tagged glucose molecules. Alternatively, an increase in gas pressure may be detected because of microbial metabolism. Another method relies upon microbes to release a fluorescent compound from the metabolism of a special precursor in the broth. The precise nature of this precursor is a commercial secret.

7.5.6   What is endocarditis and how does it develop?

Endocarditis is an infection of the lining of the heart. It typically affects one of the valves in the heart and most frequently it occurs on heart valves that have previously been damaged in some way. The most commonly affected valve is the mitral valve, then the aortic valve, followed by the tricuspid valve and finally the pneumonic valve. An array of microbes cause endocarditis; most of them are bacteria. These are often derived from the commensal flora of the patient and so endocarditis is described as an endogenous infection.

Before antimicrobial chemotherapy became available, endocarditis was invariably a fatal condition. Patients may die within a few days from acute endocarditis, such as that caused by *Staphylococcus aureus*. Alternatively, they may develop a chronic disease progressing relentlessly towards death over the course of several weeks or months. This condition used to be known as sub-acute bacterial endocarditis. The classification of endocarditis into acute and sub-acute disease is now considered obsolete.

Although many bacteria and fungi can occasionally cause endocarditis, there is a limited group of bacteria regularly found to cause the disease. The most common causes of endocarditis are the family of α-haemolytic streptococci that originally come from the commensal flora of the mouth. Heart valves that have been previously damaged by the ravages of rheumatic fever are particularly prone to endocarditis caused by these bacteria. Patients with rheumatic heart disease are at greatest risk when undergoing dental extractions. The trauma of oral surgery introduces a surge of bacteria into the bloodstream from around the tooth being extracted. These can then settle on the damaged heart valve to initiate endocarditis.

Patients with artificial replacement heart valves are most at risk of developing endocarditis caused by a coagulase-negative staphylococcus found as part of the resident flora of the skin. Most often infection is introduced at the time of the operation and if endocarditis develops within a short time then the outlook for the patient is bleak. Sometimes coagulase-negative staphylococci cause endocarditis after an artificial valve has been in place for some time. In such cases the source of the infection is not obvious. Gynaecology patients and older men may occasionally develop endocarditis caused by enterococci.

Staphylococcus aureus is a rare but important cause of endocarditis. This infection is generally associated with intravenous drug abusers who use contaminated needles. Unlike endocarditis caused by other bacteria, which typically affect valves in the left side of the heart, *Staphylococcus aureus* endocarditis attacks valves on the right side. Intravenous drug abusers inject into veins that return blood to the right side of the heart. Hence the difference in distribution. Endocarditis caused by *Staphylococcus aureus* is rapidly progressive and is fatal within days of the onset of disease. Endocarditis is also a rare complication of the rickettsial infection Q fever. This is caused by *Coxiella burnetii* although it usually manifests as a primary atypical pneumonia, this organism is a rare cause of endocarditis.

With the exception of *Staphylococcus aureus*, bacteria causing endocarditis settle on previously damaged heart valve tissue. In a healthy heart, blood flow is relatively unimpeded but if a valve is damaged then the blood flow through the heart is disturbed. There are areas around the damaged valve where the blood flow becomes slow and stagnant. This provides the initial opportunity for the bacteria to settle on the heart tissue. As they lie undisturbed, they start to grow into micro-colonies. As the micro-colonies increase in size they become covered in a fibrin deposit. This is the body's defence and ironically also its downfall. Fibrin denies other cellular defences access to the bacteria. In this protected environment, the infecting bacteria can flourish and the micro-colonies grow larger. The layers of fibrin also increase and a **vegetation** develops on the heart valve. As this enlarges, bits break off to form septic **emboli**. These can seed infections elsewhere in the body and lead, for example, to abscess formation.

Bacteria within vegetations can also stimulate the formation of circulating immune complexes. In turn these can cause further damage. A particular problem is that immune complexes can block blood flow through the kidneys, leading to renal failure. The appearance of red blood cells in urine observed under the microscope is a common observation in patients with endocarditis.

One serious consequence of the protective effect of vegetations is that the body is unable to mount an effective immune response to get rid of the causative microbes. If a patient is to recover from endocarditis then antibiotics alone must kill the infecting pathogen. Often synergistic combinations of antibiotics are used to treat the condition. A commonly used treatment is gentamicin in combination with a penicillin; therapy is maintained for six weeks. This is much longer than the average course of treatment in which the antibiotic effect is augmented by the patient's defences. During this time, serum from the patient is regularly monitored to ensure that it contains sufficient antibiotic to kill the causative bacterium.

Bacteria do not cause all cases of endocarditis. *Candida albicans* or *Aspergillus* species may cause endocarditis in susceptible patients. Viruses of the coxsackie B group can cause infection of heart muscle, referred to as **myocarditis**. This typically affects older children and young adults, who complain of chest pains and shortness of breath. Coxsackie B viruses also cause infection of the muscles of the chest. This condition is variously known as pleurodynia, epidemic myalgia or Bornholm disease, so named after an epidemic on the Island of Bornholm, lying off the coast of Denmark. This can be an extremely painful infection and patients often believe they are suffering a heart attack. A folk name given to the infection is 'The Devil's Grip'.

7.6   What causes oral cavity and respiratory infections?

The oral cavity represents the point of entry into the gut but it also lies at the entrance of the respiratory tract: 'coughs and sneezes spread diseases'. Pathogens that cause respiratory tract infections often escape from an infected individual through the mouth. It is for this reason that we have decided to discuss infections of the oral cavity here rather than with the gastrointestinal infections.

7.6.1   What causes infections of the oral cavity?

The common infections occurring within the oral cavity include dental caries, **periodontal disease** and **gingivitis**. These, together with the rather more rare actinomycosis, are caused by the activity of microbes found in the commensal flora of the oral cavity. Infections of the oral cavity may also be caused by primary pathogens. These include herpes simplex virus type 1, which causes cold sores, *Candida albicans*, the cause of oral thrush, and *Treponema pallidum*, the cause of syphilis.

7.6.2   What causes dental caries?

Dental caries attacks tooth enamel; the hardest tissue to be found in the body. The combined efforts of microbes in an accumulation of dental plaque are required to achieve a successful assault on enamel on the surface of the tooth. Dental caries tends to affect those parts of teeth that are difficult to keep clean using standard oral hygiene measures and it does not result from an attack by a single organism. Rather, it is the consequence of the activity of complex micro-communities. *Streptococcus mutans* does, however, play a central role in the initiation of a carious lesion on a tooth. There are high levels of sucrose in the Western diet and *Streptococcus mutans* can metabolise this to produce large quantities of extracellular poly-saccharides. These enable the bacteria to stick securely onto the surface of teeth. Here they produce lactic acid as a waste product of metabolism. This lowers the local pH and starts a process of demineralisation through an acid etching process.

The extracellular polysaccharides of *Streptococcus mutans* also act as an anchor for other microbes and this allows the development of the complex communities of microbes that comprise dental plaque. This is not a uniform substance and the plaque associated with dental caries is much more likely to contain high numbers of *Streptococcus mutans* than that from a healthy tooth surface. Experiments have also shown that the plaque associated with carious lesions can metabolise sucrose more efficiently than non-carious plaque. Lactobacilli and actinomycetes also play an important role in the progress of dental caries.

Although in its early stages dental caries may regress, the normal progress of the disease is relentless. First the enamel of the tooth is destroyed then dentine is attacked. Unlike tooth enamel, dentine is a vital tissue and as caries spreads to the dentine, victims suffer pain. Some people are so afraid of dental surgery that they can tolerate the intense pain of dental caries. This allows the lesion to develop and infection then spreads to the tooth pulp and through the root of the tooth culminating in the formation of an abscess. In severe cases, infection may spread from an apical abscess to cause osteomyelitis in the surrounding bone or a rapidly spreading sinus infection. It may even spread into the brain where it can initiate a brain abscess. These complications represent life-threatening conditions. The abscesses associated with untreated dental caries are invariably the result of mixed microbial infections. Anaerobic Gram-negative rods, particularly *Bacteroides* species, are commonly isolated from pus drained from these abscesses.

In the early stages of dental caries, treatment requires the radical removal of affected tissue and its replacement with one of a number of inert filling materials. Put in more simplistic terms it represents the dreaded 'drilling and filling'. For anyone who has undergone dental surgery, prevention will be better than a cure. Over recent years, a number of measures designed to prevent dental caries have been introduced. Fluorides have been added to many toothpastes and to the domestic water supplies in the UK and elsewhere. Some dentists also apply fluoride gels to the surface of teeth. This helps to harden the enamel structure making it less vulnerable to attack. An alternative approach is to use plastics to help seal the microscopic pits in the surface of teeth that act as the focus for plaque formation. In turn, this helps to prevent dental caries.

7.6.3   What is periodontal disease?

Inflammation of the gums around the margins of teeth is known as gingivitis. This is the most common form of periodontal disease: literally disease of the tissues surrounding teeth. Over 90% of the adult population of the UK suffer gingivitis to some degree. The normal flora of this anatomical region includes principally Gram-positive bacteria, particularly α-haemolytic streptococci and actinomycetes. In addition, there are also Gram-negative bacteria including spirochaetes, *Eikenella corrodens* and the anaerobic *Veillionella* and *Bacteroides* species. As gingivitis develops, the normal flora around the affected gum becomes more diverse and the number of Gram-negative bacteria increases dramatically. It is now thought that gingivitis results from a local inflammatory response to endotoxin associated with the outer membrane of Gram-negative bacteria. The gingivae then become swollen and painful. They also bleed easily. In its severest form, gingivitis may become an extensive and destructive condition, referred to as periodontitis. Young adults may occasionally suffer from an acute, severe periodontitis following surgery or accidental trauma to the gum tissue. The affected tissues are very heavily colonised by anaerobic bacteria including *Bacteroides* species and the cigar-shaped fusobacteria.

7.6.4   What is actinomycosis?

Actinomycetes constitute part of the commensal flora of the mouth. They can occasionally penetrate oral tissues to cause localised infections. Most often this follows trauma to the oral tissues such as a fracture in the jawbone. Actinomycosis results in a nodular swelling at the site of infection that becomes associated with a sinus that drains material containing small sulphur granules. *Actinomyces israelii* is the pathogen most frequently causing actinomycosis, although other species can cause the disease.

7.6.5 What is oral thrush?

Thrush is the name given to oral infections caused by *Candida albicans*. This is the most common fungal pathogen of the oral cavity. It is characterised by the appearance of numerous white plaques covering the mucous membranes of the mouth. It is very common for newborn babies to suffer a bout of thrush, but it is unusual for healthy adults to suffer unless they are taking corticosteroids or antibiotics. People who regularly inhale steroids to control the symptoms of asthma are advised to rinse their mouth after inhaling the medication to help to prevent oral thrush. Without pre-disposing factors, oral thrush in adults is generally symptomatic of an underlying disease such as diabetes or a malignancy.

7.6.6 What causes cold sores?

Herpes simplex virus type 1 causes an inapparent infection during childhood. At its worst, the primary infection with herpes simplex causes a mild fever and the appearance of cold sore vesicles around the margins of the mouth. A very high proportion of the human population across the world have been infected by the age of 10 years although only a few people show evidence of infection. Following the initial infection, herpes simplex virus travels along the nerves that serve the mouth to set up a latent infection in the trigeminal ganglion. Periodically and in response to a variety of environmental stimuli the latent infection will reactivate in a few people. The virus will replicate once more and will travel back down the nerve to cause vesicular ulcers characteristic of cold sores. Factors associated with reactivation of herpes simplex virus latency include exposure to sunlight, miscellaneous concurrent infections, emotional and mental stress and menstruation. Vesicular fluid from cold sores contains large numbers of virus particles, but these may also be shed in the saliva of asymptomatic carriers. When doting relatives kiss young children they may bestow something rather more sinister along with their affections. Herpes simplex virus type 1 may also reactivate to cause more defined herpetic ulcers within the mouth, but these are rarer than cold sores.

7.6.7 What are upper respiratory tract infections?

The human respiratory tract can conveniently be divided into two sections. The area above the vocal cords is the upper respiratory tract with its rich and diverse microflora. Below the vocal cords lies the lower respiratory tract, kept sterile in health by the action of the mucociliary escalator. A number of the bacteria sometimes found as commensals in the upper respiratory tract are able to cause endogenous infections, both of the respiratory tract and else-where. These include *Corynebacterium diphtheriae*, *Streptococcus pyogenes* and *Streptococcus pneumoniae*. All of these bacteria cause respiratory tract infections. Also found as commensals of the respiratory tract are *Haemophilus influenzae*, the cause of respiratory infections and meningitis, and *Neisseria meningitidis*, the cause of meningococcal meningitis. Upper respiratory tract infections are amongst the most common of infections in the UK and are responsible for very significant proportion of General Practitioner consultations, especially during the winter months.

7.6.8　What causes sore throats and glandular fever?

The most common causes of a sore throat are the viruses that cause the 'common cold'. There are many viruses that have been implicated as the cause of 'colds'. Amongst the most common are coronaviruses, rhinoviruses and adenoviruses. Coronaviruses are so called because they look like crowns when viewed in an electron microscope, but rhinoviruses and adenoviruses derive their names from the targets of their infections. *Rhinos* is Greek for nostril and *aden* is Greek for a gland. Coronaviruses are responsible for the majority of sore throats.

When caused by viruses, the treatment for sore throats should only be directed towards alleviating symptoms. They do not require a course of anti-biotics. Unless the patient is allergic to aspirin this is an excellent treatment for sore throats in adults. Paracetamol should be used for younger children. This is because young children who are given aspirin when suffering from influenza may develop a rare but serious liver and brain disease known as Reye's syn-drome.

Glandular fever results from an infection caused by the Epstein–Barr virus, one of the family of herpesviruses. In young children this virus can cause an asymptomatic infection. Problems arise, however, when the primary infection occurs in older children and young adults. The virus is spread by infectious saliva and the disease has been dubbed the 'kissing disease'. In a person who has not been exposed to the Epstein–Barr virus early in childhood it can cause a moderately debilitating disease of considerable duration. Symptoms typ-ically start with a patient complaining of a sore throat. This results from infec-tion of the mucous membranes in the oral cavity. Infection then spreads by way of the blood to the liver and lymph nodes. These become swollen and tender. Characteristic leukocytes known as atypical monocytes appear in the blood of people with glandular fever. These give the disease its alternate name: infectious mononucleosis. Patients with glandular fever are very lethargic and unlike other acute throat infections they do not recover rapidly. It may take several months before individuals with glandular fever return to health. Excessive exercise during convalescence may delay recovery. Patients given ampicillin to treat the symptoms of sore throat develop a characteristic skin rash that can be diagnostic for the infection.

There are now specific tests for the presence of Epstein–Barr virus but traditionally the condition was confirmed using the Paul Bunnell test. This test may take up to three months to become positive, illustrating the chronic nature of this infection. The Paul Bunnell test relies upon the demonstration of **heterophile antibodies** in the serum. These antibodies cause agglutination of sheep red blood cells. If, however, the patient's serum is first exposed to bovine red blood cells this prevents agglutination of sheep cells. Previous exposure to guinea pig red blood cells does not affect the ability of the patient's serum to agglutinate sheep red cells. Like other members of the herpesvirus family, the Epstein–Barr virus can lead to latent infection.

Epstein–Barr virus infection is associated with two cancers, Burkitt's lymphoma and nasopharyngeal carcinoma. Burkitt's lymphoma is confined principally to people who live in parts of Africa and it is thought that the disease develops in association with a cocarcinogen, probably a malaria parasite. Nasopharyngeal carcinoma is seen commonly in the population of southeast Asia and China. The likely cocarcinogen in this case is thought to be chemicals in the fish-rich diet of these peoples.

7.6.9   What causes tonsillitis?

Tonsillitis, also known as 'strep throat', is caused by *Streptococcus pyogenes* and is most common in children and young adults, although respiratory viruses are a much more common cause of tonsillar infection. Symptoms of streptococcal tonsillitis include fever, loss of appetite, difficulty in swallowing; sometimes these symptoms are accompanied by swollen glands in the neck. The tonsils become very red as they become engorged with blood, but as the infection progresses they become covered in an exudate of pus. The symptoms of streptococcal tonsillitis and the sore throat caused by virus infection cannot be distinguished on clinical grounds alone. Diagnosis requires the isolation of *Streptococcus pyogenes* from a throat swab. This is inoculated onto a fresh blood agar plate. *Streptococcus pyogenes* is a β-haemolytic streptococcus. On a fresh blood agar plate, colonies are surrounded by a clear zone of complete haemolysis. The zone of haemolysis may be enhanced if the culture is incubated anaerobically. This is because *Streptococcus pyogenes* produces two enzymes capable of destroying red blood cells. These are streptolysin O and streptolysin S. Streptolysin O is inactivated by oxygen. Penicillin is the treatment of choice for streptococcal tonsillitis although erythromycin can be used if the patient has a history of penicillin allergy.

At one time tonsillectomy, removal of the tonsils, was a routine operation performed on nearly all young children. This was in an attempt to control the rare but severe complications associated with streptococcal infection. Today the fashion for tonsillectomy has faded and the operation is no longer routine. It is not easy to remove all the tonsil tissue and any tissue remaining may regrow, thus providing a new focus for streptococcal infection.

There are two important but rare complications of streptococcal infection: **glomerulonephritis** and **rheumatic fever**. Neither condition is an infectious process: rather, they are immunological diseases. The infecting *Streptococcus pyogenes* carries antigens that cross-react with human antigens. When the patient mounts an immune response against the bacteria, antibodies are produced that also interact with human tissues carrying the cross-reacting antigens. In the case of rheumatic fever, the damage is a direct consequence of anti-streptococcal antibodies reacting with human tissue. The tissues lining the heart and heart valves are commonly affected in rheumatic fever and this is a major pre-disposing factor for the later development of endocarditis. In glomerulonephritis, circulating antibodies generate immune complexes that lodge within the glomeruli of the kidneys. This sets up a local inflammation as the complement cascade becomes activated. This leads to a progressive kidney damage and the eventual result may be total kidney failure requiring long-term dialysis or a renal transplant.

7.6.10   What is mumps?

Mumps is a generalised virus infection affecting lymphoid tissue that after about a week localises as an infection of the salivary glands. It is spread by infected air-borne droplets. The infection causes a characteristic painful swollen neck. The neck region becomes very tender while the glands are swollen. This infection generally affects children but may strike later in life. In males of reproductive age, **orchitis** is a complication of mumps infection. This is tender, painful swelling of the testes and may lead to sterility. Mumps infections may also rarely cause meningitis and encephalitis. Although there is no effective cure for mumps, it may be prevented by the use of a live, attenuated virus vaccine, now given in combination with measles and rubella.

7.6.11   What is diphtheria?

Diphtheria is caused by strains of *Corynebacterium diphtheriae* carrying a bacteriophage that codes for the diphtheria toxin. Only those strains that carry the bacteriophage are capable of causing disease since only they can produce diphtheria toxin. A long-standing vaccination programme has all but eradicated diphtheria from the UK. Sporadic cases occasionally occur and these are often seen in travellers returning from abroad. It is still endemic in other parts of the world, particularly in areas of great social, political and economic instability. In the recent past there has been a huge resurgence of the disease in the former communist countries of Eastern Europe, for example. Social upheaval disrupted the vaccine programme and this was also coupled with a fear that inadequately sterilised needles would be use for vaccination, bringing with it the risk of acquisition of AIDS or hepatitis.

Diphtheria begins with a sore throat. The patient then suffers an acute inflammatory reaction affecting the upper respiratory tract. As the diphtheria toxin causes local tissue damage there is an accumulation of a sickly smelling exudate. This leads to the formation of a false membrane, which, in severe cases, may block the airways. Local lymph nodes become congested and swollen. They fail to drain the lymph and the neck becomes swollen with accumulated fluid. This leads to a characteristic 'bull neck' appearance as fluid collects locally in the tissues. Such collections of fluid in tissues are referred to a **oedema** (*oidema* is the Greek word for a swelling). Bacteria multiply in the infected throat where they produce a powerful exotoxin that circulates around the body damaging nerve endings, the adrenal glands and tissue in the myocardium around the heart.

Primary treatment for diphtheria is concerned with maintaining a clear airway. Attempts to remove false membranes surgically are difficult since the local tissue damage may lead to a profuse bleeding, further complicating the disease. If the airway becomes blocked the patient will require a tracheotomy. This involves a surgical incision in the trachea below the obstruction in the throat. A tracheotomy tube is then introduced into the airway through which the patient may breathe. Penicillin is the treatment of choice for diphtheria but erythromycin can be used as an alternative therapy for patients with a history of penicillin allergy. Prevention is better than cure and there is a highly

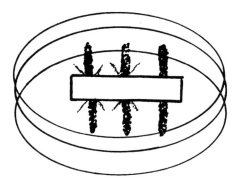

Fig. 7.4. The Elek plate. A filter paper strip soaked in diphtheria antitoxin is laid over three test strains and the plate is incubated. The right-hand strain is not toxigenic but the other two strains do produce toxin. This interacts with the antitoxin that diffuses out from the filter paper strip and the resulting antibody–antigen complex forms lines of precipitate within the medium.

effective and safe vaccine offered to infants that affords protection against diphtheria.

Laboratory diagnosis of diphtheria relies upon the isolation of the Gram-positive bacillus from clinical material and upon the demonstration of toxin formation. Selective media such as Hoyle's agar containing potassium tellurite are used for the primary isolation. *Corynebacterium diphtheriae* grows as black colonies after overnight incubation on this medium. Not all strains of *Corynebacterium diphtheriae* carry the virulence bacteriophage and consequently not all isolates can cause disease. To test for toxin production the Elek plate is used (Fig. 7.4). Strains to be tested, together with positive and negative controls are inoculated onto the Elek plate and the inocula are overlaid with a filter paper strip soaked in antitoxin. As the toxin diffuses from the growing, virulent bacteria and the antitoxin diffuses from the filter paper, lines of precipitation appear within the medium as toxin and antitoxin react. Strains that do not yield precipitin lines are considered not to produce the toxin and hence are not virulent.

7.6.12   What is acute epiglottitis?

Acute epiglottitis is now an extremely rare condition that affects very young children. It is caused by encapsulated strains of *Haemophilus influenzae*. These infect the epiglottis causing it to swell to such an extent that the airway becomes blocked. Without surgical intervention and an emergency tracheotomy the patient will suffocate. Antibiotics play an important role in the treatment of acute epiglottitis as well. It is important to eradicate the organism. If this is not done the patient may suffer a relapse. Occasionally, patients who have recovered from an acute attack may shortly afterwards develop another infection, such as meningitis, caused by the same bacterium. The choice of antibiotic used to treat acute epiglottitis can be difficult. There is a problem with the use of ampicillin. An increasing number of strains produce a β–lactamase that destroys the drug. Chloramphenicol or an extended-spectrum cephalosporin such as cefotaxime are the agents of choice. The use of chloramphenicol is, however, associated with serious side effects including **aplastic anaemia**. The widespread use of the Hib vaccine has largely eradicated acute epiglottitis from the UK. A variety of viruses, particularly the parainfluenza virus, can cause infection of the epiglottis and surrounding tissues. Clinical symptoms include fever and a harsh cough, symptoms of 'croup'. This condition sounds very frightening but it is self-limiting.

7.6.13   What causes middle ear infections?

Middle ear infections, also sometimes known as *otitis media,* are caused by a variety of agents. Patients with middle ear infections complain of an increasing earache caused by the accumulation of pus behind the eardrum. Local inflammation causes the eardrum to become markedly red from the increased blood flow in the area. As pus builds up the pain can become intense and so, too, does the pressure within the middle ear. Untreated, this pressure may cause the eardrum to rupture spontaneously. As this happens, the patient experiences an immediate relief from pain since the pressure on surrounding tissues is released.

Virus infections of the middle ear are generally mild in their symptoms. The viruses that infect tissues in the middle ear are those that are also associated with the common cold, particularly rhinoviruses and coronaviruses. Bacterial infections of the middle ear are much less common but are more severe in the symptoms they cause. Bacteria known to cause middle ear infections include *Streptococcus pneumoniae*, *Haemophilus influenzae* and *Streptococcus pyogenes*. The role of antibiotics in the treatment of middle ear infections is uncertain and treatment is aimed at relieving symptoms. Sometimes the eardrum is surgically disrupted to allow pain relief. The same pathogens that cause middle ear infections can also cause sinusitis, the painful infection of the facial sinuses on either side of the nose. Sinusitis can be very difficult to treat effectively using antibiotics.

7.6.14   What are lower respiratory tract infections?

There are a large number of microbes that can breach our defences, overcome the activity of the mucociliary escalator and establish infection within the lung. The initial diagnosis of a lower respiratory tract infection relies upon clinical examination. Listening to breath sounds and the reverberation noises made by tapping the chest can give important clues to the nature of the infection. These are often supplemented by examination of a chest X-ray. When considered in conjunction with epidemiological evidence, the pattern of inflammation or fluid consolidation in the lung will often point to the most probable causative agent. In the UK and North America pneumonia is one of the most prominent causes of death from infection.

When pathological samples from patients with lower respiratory tract infections are investigated the possibility of tuberculosis must always be excluded. This potentially fatal infection can be easily acquired by staff working in diagnostic laboratories. Consequently, all sputum samples are assumed to be positive for the presence of *Mycobacterium tuberculosis* until proved otherwise. They are also handled in special containment facilities to prevent laboratory-acquired infection.

7.6.15   What causes chronic bronchitis?

Chronic bronchitis is a disease that is caused by the interplay of a number of different factors that lead to the overproduction of mucus within the lungs. Although sputum cultures can be very difficult to interpret in cases of chronic bronchitis, a majority of patients have cultures that yield strains of *Haemophilus influenzae*. When patients with positive *Haemophilus influenzae* sputum cultures are given appropriate antibiotic therapy their symptoms do improve, suggesting that this bacterium may play a key role in the exacerbating the symptoms of chronic bronchitis.

7.6.16   What causes pneumonia?

Lobar pneumonia is an infection affecting a discrete lobe within the lung. It most commonly affects elderly patients but does occasionally occur in young adults. It is almost invariably caused by encapsulated strains of *Streptococcus pneumoniae*. Before the advent of antibiotics, the mortality associated with lobar pneumonia was very high. Patients would rapidly succumb to the disease following the sudden onset of symptoms. This led to the disease being dubbed 'the old man's friend', since it curtailed or prevented a protracted terminal illness. Because of its rapid virulence, *Streptococcus pneumoniae* has also been called 'The Captain of the Men of Death'.

Pathogenic strains of *Streptococcus pneumoniae* all produce an extracellular polysaccharide capsule and over 85 capsular types are now recognised. Bacteria that lose their capsule are no longer capable of causing disease. Consequently, the capsule is an important virulence factor. A vaccine has been produced to protect against infection caused by the 23 most common capsular types of *Streptococcus pneumoniae*. Vaccination is offered to those people who are particularly susceptible to pneumococcal infection. Ironically, those who are most vulnerable to infections caused by *Streptococcus pneumoniae* are the least likely to derive the full benefit from vaccination.

A heavy growth of *Streptococcus pneumoniae* may be diagnostic when isolated from the sputum of a patient with suspected lobar pneumonia. These bacteria are, however, found as commensals in the upper respiratory tract of a significant number of healthy people. As such they may easily contaminate a sputum sample on its passage from lung to specimen pot. Indeed, most cases of pneumococcal pneumonia are endogenous since they arise from the patient's own flora. In about half of cases, lobar pneumonia is accompanied by a bacteraemia. Unlike sputum samples, blood culture does not rely upon passage of the clinical sample through an area densely populated with a microflora and *Streptococcus pneumoniae* is not a skin commensal. Isolation of pneumococci from the blood culture is very valuable in confirming the diagnosis of a patient with suspected lobar pneumonia.

The treatment of choice for pneumococcal pneumonia is penicillin, although erythromycin can be used in patients with a history of penicillin allergy. This treatment may, however, be under threat. Pneumococci have emerged that have altered penicillin-binding proteins. These bacteria are relatively resistant to the normal therapeutic doses of penicillin, and pneumonia caused by these strains can only be successfully treated if very high drug doses are used. Penicillin-tolerant strains of pneumococci represent a major problem if the strain in question goes on to cause infection at a site where antibiotics have poor penetration, such as in meningitis.

Bronchopneumonia is an infection that is widely dispersed throughout the lungs. Typically, numerous small areas of consolidation are to be found across the lungs. There are many causes of bronchopneumonia. A rare but important cause of bronchopneumonia is *Staphylococcus aureus*. This bacterium typically causes a secondary infection of the lungs, exploiting damage caused by a primary virus infection such as influenza. Staphylococcal pneumonia is diagnosed in the laboratory by observing typical clusters of Gram-positive cocci in sputum and the subsequent heavy growth of *Staphylococcus aureus* upon culture of the specimen. The sputum of patients with staphylococcal pneumonia is heavily laden with pus. Consequently, it takes on a creamy appearance and is also often stained with blood.

Alcoholics are prone to developing aspiration pneumonia. This is a relatively rare condition that results from the inhalation of vomit, acting as a focus for subsequent infection. Coliform bacteria such as *Klebsiella pneumoniae* frequently cause aspiration pneumonia and they, together with *Pseudomonas aeruginosa,* cause bronchopneumonia in patients who require assisted ventilation to support their breathing. Such patients are commonly to be found in intensive therapy units.

Children and young adults with cystic fibrosis are prone to lung infections. Cystic fibrosis is associated with the production of a sticky, glutinous secretion within the lungs that pre-disposes to infection. Early bouts of pneumonia are caused by *Staphylococcus aureus*. As the child grows older, *Haemophilus influenzae* is a more frequently isolated pathogen. Later still, *Pseudomonas aeruginosa* is a common cause of pneumonia. Patients with cystic fibrosis are given repeated courses of antibiotics to tackle these episodes and this helps to contribute to the selection of increasingly resistant pathogens. Ultimately patients with cystic fibrosis will suffer infections caused by unusual and highly antibiotic-resistant pseudomonads such as *Burkholderia cepacia*. These bacteria are highly mucoid, which contributes to the disease process. They are also impossible to eradicate and are a frequent cause of death in patients with cystic fibrosis.

Atypical pneumonia is a loose term applied to cases of pneumonia that are not caused by the 'classical' bacterial pathogens. Such cases generally have an gradual onset and are associated with a dry, unproductive cough. There is often little or no evidence of consolidation apparent within the lung tissue.

Mycoplasma pneumoniae can cause a severe primary pneumonia that has an insidious onset. Because these bacteria are difficult to grow in artificial culture, the laboratory diagnosis depends upon serological evidence: detecting antibodies in a complement fixation test. Mycoplasmas also lack a peptidoglycan-based cell wall. Consequently they are not susceptible to treatment with penicillin. Erythromycin is considered to be the best therapeutic option for pneumonia caused by mycoplasmas since it is also effective against mis-diagnosed cases of Legionnaire's disease. Although tetracycline can be used to treat pneumonia caused by *Mycoplasma pneumoniae*, it is ineffective against *Legionella pneumophila*.

Coxiella burnetii is the cause of a rare atypical pneumonia referred to as

Q fever. This is a highly fastidious obligate intracellular parasite and it can only be artificially cultivated in tissue cultures. Diagnosis of *Coxiella burnetii* pneumonia relies on a complement fixation test. Antimicrobial therapy exploits tetracycline or chloramphenicol. Bird fanciers who keep parrots are prone to develop psittacosis (*Psittakos* is the Greek word for a parrot). This is a serious pneumonia that is susceptible to treatment with tetracycline. The causative bacterium is *Chlamydia psittaci*: another obligate intracellular parasite.

Pneumocystis carinii is a member of the commensal flora of the upper respiratory tract. For many years its microscopic appearance and lifestyle caused this organism to be classified as a protozoan, but recent molecular studies have shown it to be a highly atypical fungus. In severely immunocompromised patients it can cause pneumonia. *Pneumocystis carinii* pneumonia is often the clinical condition that marks the transition to full-blown AIDS in patients who are HIV positive. Chest X-rays show that both lungs are affected in the lower portion, with a diffuse shadowing over the affected areas. Patients suffer a dry unproductive cough and shortness of breath. They may have such difficulty in breathing that they appear blue owing to a lack of oxygenated blood. The diagnosis of *Pneumocystis carinii* pneumonia depends upon a bronchoscopic examination and retrieval of the characteristic cysts from bronchial washings. The condition is difficult to treat but may be controlled with a combination of trimethoprim and a sulphonamide.

Influenza is the best known of virus pneumonias. The influenza A virus undergoes continual genetic modification and as a result people can suffer from several bouts of influenza. The severity of each attack depends upon the nature of the genetic change in the virus. Small changes in the virus structure resulting from the accumulation of point mutations in the virus genome are referred to as **genetic drift**. People who have suffered a bout of influenza and who are subsequently infected by a strain that has undergone genetic drift will suffer a milder disease course than upon the first encounter. This is because the immune response provoked on the initial infection will provide partial protection from the second infection. Influenza virus is, however, genetically unstable. The RNA genome of this virus is segmented. Influenza virus can infect a number of animals and birds and different strains of the virus become adapted to their particular host species. Very occasionally, humans, animals or birds may be simultaneously infected with two separate strains of influenza virus. One virus strain is normally associated with the infected host and the other is adapted to life in a different species. This allows the opportunity for a major re-assortment of the virus genome with different RNA molecules coming randomly from either parent virus. The newly generated virus will contain major antigens never previously encountered by the human population. These new viruses can, nonetheless, infect humans. The major genetic rearrangements responsible are referred to as **genetic shift**. The new variant virus resulting from a genetic shift event will not have been encountered before and the world population will have no antibody protection. Consequently, the new virus will spread rapidly around the world causing a pandemic. Since the sixteenth century, there have been over 30 influenza pandemics. The influenza pandemic of 1918 killed 20 million people: more than died as a result of all the battles of the First World War.

Influenza is spread by the inhalation of infected droplets. Most cases are relatively trivial but in some people the infection progresses to cause bronchopneumonia. This carries a significantly increased risk of secondary bacterial infection. This is also associated with a significantly increased risk of death. The acute infection is treated symptomatically. Aspirin is recommended for adults. There is a significant risk that children with influenza may develop the liver and brain disease known as Reye's syndrome if they are given aspirin and so the recommended treatment for children is paracetamol. In vulnerable people, amantidine may be used both to treat and to prevent infection.

In recent years, vaccination programmes designed to protect vulnerable individuals from influenza has met with some success. As the influenza season approaches, a cocktail of viruses is prepared. The precise content of the vaccine depends upon an educated guess about the antigens likely to be prevalent in the forthcoming influenza season. These educated guesses are based upon global epidemiological studies.

Respiratory syncytial virus causes a mild flu-like illness in adults and older children. In infants, however, this virus causes a severe infection of the bronchioles known as bronchiolitis. This is a life-threatening infection for babies, prevalent during the winter months. Because the respiratory syncytial virus also undergoes antigenic variation, babies may suffer more than one episode of bronchiolitis in their early months. Ribavarin may successfully treat severe cases.

7.6.17 What is Legionnaire's disease?

Legionnaire's disease was first recognised when it caused a cluster of atypical pneumonia cases in delegates attending the 58th convention of the Pennsylvania Branch of the American Legion, held in Philadelphia in 1976. The press then nicknamed the disease 'Philly Killer'. Retrospective epidemiological investigation showed that this was not the first occurrence of the disease. Indeed, it was not even the first occurrence of the disease at the hotel used by the Legionnaires. Two years earlier it struck a delegation from the Independent Order of Odd Fellows. Perhaps we should be grateful that it was the American Legion incident that brought the disease to the world's attention and it was them rather than the Odd Fellows that gave the disease its name.

Legionnaire's disease is caused by the Gram-negative bacillus *Legionella pneumophila*. This species causes the most severe form of the disease although other members of the genus and of the related genus *Tatlockia* can cause milder forms of pneumonia such as Pontiac fever. Although *Legionella pneumophila* can be cultured in the laboratory, diagnosis typically relies upon serology. An infected patient will show a rise in antibodies specific for the bacterium. *Legionella pneumophila* is fastidious in its growth requirements and needs high iron levels and cysteine if it is to be grown on laboratory media.

Clinically, Legionnaire's disease resembles influenza, particularly in the early stages. Examination of the X-ray film of a patient with the disease, however, shows profound differences. The typical X-ray picture of a patient with Legionnaire's disease resembles a 'white-out' with both lungs heavily infiltrated with fluid. Patients with Legionnaire's disease will often require artificial mechanical ventilation to support their breathing. Complications of Legionnaire's disease include altered mental state and neurological symptoms. Patients may also develop renal failure and may sustain liver damage.

Legionella pneumophila is widely distributed in natural waters. It can be easily isolated from badly maintained air-conditioning systems that rely upon water-cooled heat-exchange mechanisms and from showerheads that are not regularly cleaned. Infected droplets are easily spread and can become aerially dispersed from such sources. The main route of infection is by inhalation of infected aerosol droplets. Smokers are at particular risk of developing Legionnaire's disease. The condition is also more common amongst the middle aged.

Legionella pneumophila produces a β-lactamase that renders treatment with a penicillin useless. Patients who have been mis-diagnosed as having pneumococcal pneumonia will fail to respond to penicillin therapy. Erythromycin is the treatment of choice for Legionnaire's disease and for best results it is prescribed together with rifampicin.

7.6.18   What is tuberculosis?

Tuberculosis, TB for short, is primarily a respiratory infection caused by *Mycobacterium tuberculosis*, although it may occasionally be caused by

Mycobacterium bovis. In most cases the infection is confined to the lung although lesions in the lung can seed infection elsewhere in the body. *Mycobacterium tuberculosis* can infect the kidney to cause renal tuberculosis. It is also a cause of chronic meningitis and can cause osteomyelitis typically in the bones of the spine. It may also cause a more generally dispersed infection throughout the body referred to as miliary tuberculosis. Immunocompromised patients, especially those with full-blown AIDS, are susceptible to a TB-like illness caused by the so-called 'atypical' mycobacteria, particularly those of the *Mycobacterium avium-intracellulare* complex.

Mycobacteria are obligate aerobic bacteria. Consequently they tend to cause lesions in the upper lobe of a lung where the concentration of oxygen is highest. Mycobacteria are intracellular pathogens. They can resist being killed after phagocytes have engulfed them. Indeed, once inside an alveolar macrophage, cells of *Mycobacterium tuberculosis* can grow and multiply. This causes the macrophage to die, releasing bacterial cells that can infect further macrophages, setting up a cycle of infection and multiplication. Initial infection provokes a cell-mediated immune response that contains infections within defined areas known as tubercles.

Following primary infection, tuberculous lesions become inactive and may remain in a quiescent (resting) state for many years. In later life, as our innate host defences wane, old TB lesions may reactivate causing a second phase of infectious disease. Before the Second World War, tuberculosis was a major problem. Following the war, however, there was a massive drive to control the disease through public health measures, screening and an active vaccination programme. Vaccination uses the live, attenuated bacterium known as the Bacille–Calmette–Guérin (BCG), named after the scientists who developed the vaccine. This programme was so successful that the disease went into a dramatic decline and old TB sanatoria were closed. Unfortunately this led to a degree of complacency and in recent years vaccination programmes have become somewhat lax. The problem is complicated by a fierce debate concerning the efficacy or otherwise of BCG vaccination. We are now seeing the consequences of our inaction. Tuberculosis is a newly emerging disease in economically developed countries. Grandparents infected many years ago when the disease was rife can act as a potential pool of infection for their unvaccinated grandchildren. Perhaps we can anticipate a time when tuberculosis is a major killer in the developed world just as it has remained one of the principal causes of death in the Third World. Some people are already discussing the possible imminent return of the 'White Death'.

A presumptive laboratory diagnosis of tuberculosis can be made by observing acid alcohol-fast bacilli within a sputum sample. The waxy cell walls of mycobacteria make them difficult to stain using conventional methods. Once stained, however, they retain dyes even through the harsh decolorisation protocols of procedures such as the Ziehl Neelsen or Auramine–Rhodamine methods. Mycobacteria are very slow growing and may take up to eight weeks to grow on egg-based media. They develop into colonies reminiscent of small breadcrumbs. PCR technology is revolutionising the laboratory diagnosis of tuberculosis. By amplifying DNA sequences that are specific to mycobacteria, diagnosis can be made within a day rather than after two months.

Antituberculous therapy has to be prolonged because the organism is so slow to grow. The minimum period of treatment is six months, and antituberculous drugs are often not well tolerated. Patients may, therefore, have to change to a less satisfactory course of treatment that also lasts longer. Because of the problem of emerging resistance, antituberculous drugs are always given in combination. The pattern of prescribing will be dictated by the local resistance patterns. Drugs used for the treatment of tuberculosis include isoniazid, rifampicin, pyrazinamide, ethambutol and streptomycin. Because of the extended course of therapy and the unpleasant side effects associated with treatment, patients not infrequently fail to take their medication regularly. This has led to the emergence of multiply drug-resistant *Mycobacterium tuberculosis* strains. Indeed, the problem is so bad that the health authorities in certain American cities pay patients to have supervised delivery of medication in an attempt to improve compliance rates.

7.6.19 What causes whooping cough?

Whooping cough is the infection caused by *Bordetella pertussis*. Infecting bacteria stick to the surface of the trachea and the bronchi where they release a toxin. This interferes with the normal regular beating of the cilia on the epithelial surface. Since we rely upon the beating of cilia to remove mucus from the lung, the toxin causes an accumulation of mucus. The infection also causes overproduction of mucus, further complicating the condition. Mucus accumulates to the point where the victim responds with a severe coughing fit. Often the cough ends with a characteristic 'whoop', from which the disease derives its name. A number of viruses can produce symptoms that mimic whooping cough. Important examples include the parainfluenza virus and certain adenoviruses.

In older children, whooping cough is relatively mild but in babies less than six months old it is a life-threatening condition. This is because babies have a relatively small lung capacity and their lungs are more easily overwhelmed by infection. Mucus accumulation easily causes respiratory obstruction in tiny lungs and this will consequently lead to death. Antibiotics have no useful role in the management of a patient with whooping cough. To protect infants from whooping cough a vaccine is widely used. This works by conferring 'herd immunity'. Vaccination of the vast majority of six-month-old children dramatically reduces the pool of individuals who are susceptible to the disease. This reduces the opportunity of the pathogen to spread and, in turn, this means that fewer pathogens can circulate in the community. It is in this way that children who are too young to receive the vaccine themselves can be protected from the disease.

7.6.20   What is aspergillosis?

Aspergillosis most frequently affects the lungs and may manifest a variety of clinical conditions. Extrapulmonary infections are occasionally seen, particularly in the nasal sinuses. Patients may suffer an allergic reaction to aspergilli. Alternatively, aspergilli can colonise pre-existing cavities within the lung to cause a fungus ball or aspergilloma. Rarely these fungi can cause a more invasive disease when infection becomes disseminated. This typically occurs in severely immunocompromised patients. The most important species of the genus *Aspergillus* associated with human infection are *Aspergillus fumigatus*, *Aspergillus niger*, *Aspergillus flavus*, *Aspergillus terreus* and *Aspergillus nidulans*. All grow as mycelial fungi with septate hyphae and may be identified by their characteristic sporing structures.

People with elevated IgE levels tend to exhibit allergy to a variety of stimuli. About 10–20% of asthma patients have an allergic response to *Aspergillus fumigatus*. This may provoke an increase in the number of eosinophilic leukocytes and lead to a chronic form of aspergillosis, causing progressive lung damage. Corticosteroids are used to treat allergic aspergillosis.

Members of the genus *Aspergillus* can colonise pre-existing lesions within the lung to form an aspergilloma. Old TB lesions are a favoured site for their development. A dense fibrous wall surrounds the ball of mycelia and most lesions form discrete entities. Occasionally, an aspergilloma may invade surrounding tissue and seed infection of other organs. Haemorrhage resulting from the invasion of a local blood vessel is one of the serious complications of an invasive aspergilloma. Fungus balls are usually solitary and can vary in size. Typically, however, they are less than 8 centimetres in diameter. People who have an aspergilloma may be without symptoms or may have a moderate cough that produces sputum. This may be stained with blood when the fungus is actively growing. Surgical removal is the most common and effective treatment of this condition.

7.7 What causes gastrointestinal infections?

Many gastrointestinal infections are acquired either by eating contaminated food or by drinking contaminated water. Consequently, these infections have already been discussed in Sections 3.1 and 5.6. Table 7.4 lists the features of the principal bacterial forms of gastroenteritis.

7.7.1 What is pseudomembranous colitis?

Since the early clinical application of antimicrobial chemotherapy, the use of antibiotics has been associated with diarrhoeal disease. In the case of erythromycin this appears to be a pharmacological effect of the drug on the gut and the condition is relatively mild, resolving when the patient stops taking the medication. The diarrhoeal side effects of antimicrobial chemotherapy may, however, be life-threatening. The use of clindamycin, in particular, can cause a condition known as pseudomembranous colitis, although almost any antibiotic has the potential to cause this condition. The patient develops a severe diarrhoea, passing blood and mucus in large quantities *per rectum*. Upon endoscopic investigation, the bowel is coated in a pseudomembrane. This condition is caused by an overgrowth of *Clostridium difficile* in the bowel and it carries a significant risk of mortality, particularly in elderly or frail patients. This is an obligate anaerobic spore-forming Gram-positive bacillus. It is normally kept at bay by the commensal flora of the bowel. When a patient takes clindamycin or other antibiotic, the balance of the bowel flora is disturbed, allowing *Clostridium difficile* to flourish. If the victim is unfortunate enough to have acquired a toxigenic strain then pseudomembranous colitis results.

There are two toxins associated with *Clostridium difficile*. One is an enterotoxin and the other is a cytotoxin. Both play an important role in causing severe diarrhoea. Although it is caused by use of an antibiotic, it may also be cured by antibiotics. Metronidazole is the treatment of choice but oral vancomycin can also be used. Vancomycin is not absorbed from the gut and for all other uses should be delivered **parenterally**. Despite apparently successful antibiotic therapy, the disease can recur in a significant number of patients. At one time this was thought to be the result of recurrence of the original

Table 7.4 *The principal symptoms of some of the major causes of bacterial gastroenteritis*

Causitive agent	Incubation period	Duration of illness	Diarrhoea	Vomiting	Abdominal cramps	Fever
Campylobacter species	15 hours to 2 weeks	3 days to 3 weeks	+++	+/−	++	++
Salmonella species	15 hours to 2 days	2–7 days	++	+	+	+
Shigella species	1–4 days	2–3 days	++/+++	−	+	+
Vibrio cholerae	2–3 days	Up to 7 days	++++	+	−	−
Vibrio parahaemolyticus	6 hours to 2 days	3 days	+/++	+	+/++	+
Clostridium perfringens	8–24 hours	12–24 hours	++	−	++	−
Bacillus cereus	8–12 hours	12–24 hours	++	−	++	−
Bacillus cereus	30 minutes to 6 hours	12–24 hours	−/+	++	−	−
Staphylococcus aureus	30 minutes to 6 hours	12–24 hours	−/+	++++	−	+
Yersinia enterocolitica	4–7 days	12–24 hours	++	−	++	+

Note: Absence of symptoms is indicated by a minus sign (−). The range of severity of symptoms in cases of bacterial gastroenteritis is indicated by increasing numbers of plus signs, with + being mild symptoms up to ++++ being very severe symptoms.

infection but recent studies have shown that in many cases the recurrence is caused by a different strain. This indicates that the patient has become re-infected, possibly from a different source, rather than simply succumbing to a relapse of the first infection.

7.7.2 How are faecal samples examined for pathogens?

In the routine examination of faeces, the commensal microorganisms within the gut may vastly outnumber pathogens. Laboratories cannot rely upon primary culture techniques to isolate faecal pathogens. For this reason, a secondary selective process is used. The choice of media used vary from laboratory to laboratory and to some extent are determined by the preferences of the local microbiologist.

At the University of Leeds we use a xylose lysine deoxycholate (XLD) plate and a deoxycholate citrate agar (DCA) plate together with a selenite enrichment broth for the routine examination of faeces. These are inoculated and the primary plates examined for the presence of pathogens after incubation for 24 hours. If none are seen, the primary plates are incubated for a further 24 hours. After the first day the selenite broth is sub-cultured to further XLD and DCA plates and these are incubated for 24 hours.

To isolate campylobacters, a VPAT plate is inoculated and incubated microaerophilically for 48 hours at 43°C. This is nutritionally rich medium that contains vancomycin, polymyxin, amphotericin B and trimethoprim and is used to inhibit the growth of faecal commensals. If cholera is suspected or if the stool has a 'rice-water' appearance, the sample is plated onto a thiosulphate citrate bile salts sucrose (TCBS) agar plate. An alkaline peptone water is also inoculated to serve as an enrichment medium, since *Vibrio cholerae* is tolerant of alkaline conditions. The alkaline peptone water is sub-cultured onto a further TCBS plate after incubation.

The laboratory diagnosis of a patient with suspected haemolytic-uraemic syndrome is complicated because it is caused by *Escherichia coli*. Fortunately, *Escherichia coli* O157 fails to metabolise sorbitol, whereas the commensal strains can do so. To identify the causative bacterium a MacConkey plate in which lactose is replaced by sorbitol will indicate the presence of the pathogen. Tests for other enteropathogenic strains of *Escherichia coli* may be performed without the need for sub-culture from selective medium. Stool samples from children under 2 years of age are plated onto fresh blood agar and serological tests are used on the primary isolates.

Occasionally, *Bacillus cereus* infection is investigated. This is done by pasteurising the stool sample to kill the vegetative microbes in the sample then plating it onto fresh blood agar incubated aerobically. *Clostridium difficile* is cultivated on a medium containing cycloserine and cefoxitin incubated anaerobically for 48 hours. This bacterium shows a yellow-green fluorescence when viewed under long-wave ultraviolet light. Not all isolates are toxigenic and the presence of cytotoxin must be demonstrated if an isolate is to be considered pathogenic. This is usually done with either a cytotoxin assay using tissue culture cells or by an ELISA test. The faeces of young children with suspected rotavirus infection are subjected to latex agglutination tests to detect the virus.

7.7.3 What viruses are associated with gastroenteritis?

Viruses are the most common causes of gastrointestinal infection worldwide. Rotaviruses alone cause 20% of cases of diarrhoea in developing countries. Many cases occur in infants, and virus gastroenteritis is responsible for significant infant mortality in the Third World. In developed countries, up to 60% of infants who need hospital admission to treat the dehydration caused by diarrhoea are suffering from one of the virus enteritides. Breast-fed babies are less likely than are bottle-fed babies to develop virus gastroenteritis. This is taken as evidence that breast milk has a protective effect, stopping infants from acquiring virus diarrhoea.

Although rotaviruses are the most common cause of virus enteritis, a number of other viruses have also been implicated in this disease. These include adenoviruses, caliciviruses, astroviruses and the small round viruses associated with eating shellfish. The caliciviruses are typically associated with infections in older children and adults. They also show a marked seasonal variation, being the cause of 'winter vomiting disease'. This can strike dramatically. One of the authors of this book was attending a school Christmas concert when, in the middle of a carol, a child in the back row of the chorus developed the symptoms of this infection. This precipitated what the conductor referred to as 'a sawdust incident' but not before the front row of the choir received a thorough dowsing.

The most notorious of the caliciviruses causing enteritis is the Norwalk agent. This was named after the school in Norwalk, Ohio where it was first isolated. It causes myalgia and headache as well as nausea, vomiting and diarrhoea but the symptoms generally resolve within a day. This has been studied in volunteer trials to illustrate its pathogenicity. Virus enteritis is not confined to children. Large numbers of adults can become infected simultaneously, especially on cruise ships or in hospital wards. The infectious dose can be very low indeed. Fewer than ten virus particles can precipitate infection. This can make the job of disinfection very hard and certain cruise ships seem to be a chronic cause of virus enteritis. The viruses responsible for enteritis are transmitted by the faecal–oral route.

Most of the viruses that cause gastroenteritis cannot be propagated in tissue culture although they can often be seen in electron microscopy. The cost of this procedure means that it is not used for routine diagnosis of diarrhoeal disease. Latex agglutination tests may be used but most often the diagnosis is made by excluding other causes and by examining the balance of probability.

7.7.4 What causes hepatitis?

Many viruses, for example the Epstein–Barr virus, yellow fever virus and others, can cause hepatitis. When considering hepatitis, however, we normally refer simply to the hepatitis viruses: the major cause of virus hepatitis. Until recently there were three recognised types of virus hepatitis: hepatitis A, hepatitis B and non-A non-B hepatitis. This changed in 1989 with the first characterisation of the hepatitis C virus and now hepatitis D, hepatitis E, hepatitis F and hepatitis G have each been described. There are probably other viruses that cause hepatitis yet to be described.

Hepatitis A is a picornavirus, a small virus with a single-stranded RNA genome. At one time it was classified with the enteroviruses but it has recently been separated taxonomically. It is spread by the faecal–oral route and has been discussed in Section 3.1.5. Hepatitis B is a double-stranded DNA virus present in the blood of infected people. Intravenous drug users who share needles to inject their drugs are at grave risk of acquiring hepatitis B infection, as are people whose sexual practices are more likely to cause bleeding than others. People who have tattoos or who engage in body piercing are also at risk unless scrupulous hygiene is observed during the operations. Occasionally, the use of acupuncture needles have been implicated in the spread of hepatitis B. Although virus particles can reach very high numbers in the blood of infected persons, blood-sucking insects have not been shown to spread hepatitis B. Blood transfusions were an important vector before screening was widespread.

The surface antigen of the hepatitis B virus is used as a marker of infection. At one time this antigen was widely known as the Australia antigen. In a study of the genetics of human blood types, sera from haemophiliacs were used to react with blood samples taken from representatives of a range of human races. It was found that haemophiliac serum always reacted with an antigen in the blood of Australian Aborigines. It is now known that Aborigines are very likely to be asymptomatic carriers of hepatitis B and this is the source of the Australia antigen. This is now more often referred to as the HBsAg, short for hepatitis B surface antigen. Before widespread screening, haemophiliacs acquired antibodies to this virus through repeated transfusion with infected blood products: hence the reaction. There are other important antigens associated with the hepatitis B virus. The HBcAg is associated with the core of the virus and the HBeAg is a marker of chronic infection. If this antigen is detectable in a blood sample then the patient is very likely to develop cirrhosis of the liver.

After infection, there is a long incubation period. This may last from several weeks to a few months. In many cases the infection never exhibits clinical signs. In the UK approximately one in a thousand people is a chronic, asymptomatic carrier of the hepatitis B virus. A major problem is that asymptomatically infected mothers may pass hepatitis B infections vertically to their offspring, who have a much higher risk of becoming asymptomatic carriers. Worldwide there are an estimated 300 million carriers of the hepatitis B virus. If symptoms develop then jaundice is not the first sign of disease. Patients first develop a skin rash and suffer inflammation of their joints. This is referred to as the pre-icteric phase of the disease. Only after these clinical signs does the patient become jaundiced. The primary liver damage is much more severe than is seen in hepatitis A and can last for several months. In some cases, the infection becomes chronic, leading eventually to cirrhosis of the liver. Hepatitis B carriers are more than 200 times more likely than other people to develop carcinoma of the liver. This develops many years after the initial infection. Although there is no active treatment for hepatitis B there is an effective vaccine. This has been made through recombinant DNA technology. The active epitope of the surface antigen has been cloned. This is expressed to form a subunit vaccine, so called because its use does not require exposure to the entire virus particle. The hepatitis B subunit vaccine is offered to all those whose work brings them into contact with human blood or blood products.

The hepatitis C virus is an RNA virus that cannot be grown in artificial culture. Its genome nucleotide sequence has been determined using PCR coupled with a preliminary reverse transcriptase reaction to convert the virus RNA into a DNA template. Using this PCR technology it has been shown that about 1% of people in developed countries carry the hepatitis C virus. It typically causes hepatitis following transfusion of infected blood. Its mode of transmission is essentially the same as the hepatitis B virus. It is this virus that causes the condition once known as non-A non-B hepatitis. The virus causes active chronic hepatitis in about 80% its victims and of these about one fifth will develop cirrhosis because of their infection. Infection with the hepatitis C virus, like the hepatitis B virus, carries an increased risk of developing carcinoma of the liver. Worldwide there are at present an estimated 200 million carriers of the hepatitis C virus.

The hepatitis D virus is a small RNA virus that is also a defective virus. It can only cause infection in patients who are already infected with the hepatitis B virus. About 5% of people infected with hepatitis B are also infected with hepatitis D. This virus is transmitted in the same way as the hepatitis B virus and can be found in very high numbers in the blood of infected people.

The hepatitis E virus is a small single-stranded RNA virus that is transmitted in a similar way to the hepatitis A virus, although its incubation period is generally longer. Most often the infection is mild and self-limiting but in pregnant women it can cause a life-threatening condition involving generalised organ failure. The hepatitis F and G viruses are assumed to cause blood transfusion-associated hepatitis with no other attributable cause. The role of these viruses in causing disease is uncertain at present.

Besides virus hepatitis, bacteria may infect the gall bladder. This occurs most often when gallstones obstruct the bile duct. In this case, the most likely causative bacterium is one of the Enterobacteriaceae, especially *Escherichia coli*, although anaerobes such as *Bacteroides fragilis* have also been implicated. The gall bladder of typhoid carriers is also often colonised and may be the source of chronic infection. Antibiotic therapy is not successful in treating gall bladder infections: most antibiotics do not function well in the presence of bile. Surgical removal of the infected gall bladder may be the only sure way to cure such an infection.

7.7.5 What is peritonitis?

The lower gut houses an immense microflora. All the time it is retained within the gut there is no problem but should the gut contents leak into the peritoneal cavity then a life-threatening infection results. Leakage of gut contents may occur through accidental trauma or surgery and is also the consequence of the rupture of an acutely inflamed appendix. Peritonitis caused by leakage of gut contents is a mixed infection with both coliform bacteria and the obligate anaerobic *Bacteroides* species being important pathogens. If abscesses form within the peritoneal cavity, these may require surgical drainage. Antibiotic therapy should include metronidazole to treat the obligate anaerobic bacteria as well as appropriate cover for the facultative organisms.

A less acute peritonitis is associated with patients undergoing continuous ambulatory peritoneal dialysis (CAPD). This is a treatment for renal failure and is an alternative to using dialysis machines. Dialysis fluid is introduced into the peritoneal cavity through a permanent catheter and is left in place to accumulate waste products. After an appropriate period the fluid containing waste products is drained off and replaced with fresh dialysis fluid. If the catheter becomes infected it may seed infection in the peritoneal cavity. The bacteria most commonly associated with this type of infection are the coagulase-negative staphylococci from the patient's skin. The risk of developing peritonitis associated with CAPD is high. Patients change dialysis fluid between four and six times per day. In a year, a sterile environment will potentially be breached on at least 1500 occasions. If peritonitis develops, the catheter must be removed if the infection is to be cured, even with appropriate antimicrobial therapy.

7.8 What causes infections of skin, bone and soft tissues?

Skin provides an inhospitable mechanical barrier, forming one of our important first-line defences against infection. Despite this, there are fungi, viruses and bacteria that can establish infections of the skin. Furthermore, several systemic virus infections become clinically apparent because they cause skin rashes. In many cases skin infections develop following trauma. Through breaches in our skin, pathogens may gain access to our sub-cutaneous tissues that are kept sterile in health. Infection of these tissues can have disastrous consequences. Alternatively, our bones and soft tissues may become infected by way of the bloodstream from colonisation or infection at another site in the body.

7.8.1 What bacteria cause skin and muscle infections?

Healthy skin rarely becomes infected. Skin infection generally occurs because of trauma breaking the integrity of the epidermis, allowing pathogens to colonise the damaged tissue. Sometimes the damage can be so slight as to be inapparent. Although many of the skin traumas that lead to infection are accidental, surgical wounds are made deliberately. These are prone to infection although the risk varies with the site of the wound and the nature of the operative procedure. Clean wounds giving access to sites that are sterile carry a slight risk of infection: typically less than 1% become infected. If the operative site is associated with a commensal flora then the risk of developing a wound infection increases to about 1% for most sites. Such sites are sometimes called 'clean contaminated' sites. The risk of infection can rise as high as 15–20% for bowel surgery since this carries a heavy microbial load. Although the gut flora does not cause problems when contained, many constituents can cause life-threatening infections when they gain access to tissues beyond the gut. Control of surgical wounds requires careful attention to surgical procedure and good post-operative care. Surgical wounds may often become infected with *Staphylococcus aureus* because of poor post-operative procedures. To prevent infections in 'dirty' surgical wounds, prophylactic antibiotics given for a short time around the period of the operation may provide protection from subsequent infection.

Skin infections can be very unsightly and can cause immense psychological damage. They have even been directly responsible for suicide. In the past, fashions in clothes have evolved to hide skin lesions. Beauty marks popular in the seventeenth century hid the skin lesions associated with syphilis and in the eighteenth century high collars were worn to disguise the lesions associated with scrofula, a chronic mycobacterial infection. Today, adolescents spend vast sums of money on proprietary products used to control the ravages of acne. This disease is associated with the microaerophilic bacterium *Propionibacterium acnes*. Although much research has been carried out concerning the precise role of this bacterium in acne, it is still not clear what role it has to play in the disease. People without acne carry the bacterium as a commensal in the skin and people with acne may not yield *Propionibacterium acnes* from their lesions. There is, however, considerable evidence to suggest that this bacterium does play a central role in the pathogenesis of acne, a multifactorial disease influenced by bacteriological, immunological and hormonal influences. Certainly, long-term antibiotic therapy with tetracyclines or erythromycin can cause relief of symptoms in most cases of acne. As there are alternative effective treatments for acne, antibiotic therapy should be reserved for the most severe or intractable cases.

Staphylococcus aureus is an important bacterium causing a range of skin infections. Some people carry this bacterium on the moist areas of skin in the armpit or in the groin area. In its least aggressive form it appears harmless. *Staphylococcus aureus* infects hair follicles or sebaceous glands to cause boils, more properly referred to as **furuncles**. Initially, the affected area becomes very red, hard and swollen. The infected tissue and its surroundings become tender to touch. A large quantity of yellow pus collects within the infected site and erupts as the boil 'comes to a head'. A boil results from the infection of a single follicle. If a group of follicles become infected, then a crop of boils in one location will develop. These are collectively known as **carbuncles**. Crops of pustules caused by *Staphylococcus aureus* and appearing on shaved skin is referred to as 'barber's rash'. In such cases microscopic damage to the skin occurs during the shaving process and this allows access to the pathogen. Barber's rash is a particular problem for people who wet shave with cut-throat razors. These are noted for the closeness of the shave. *Staphylococcus aureus* may also cause infection of the fingernails or nail bed, causing a **whitlow**. Again, this is characterised by an accumulation of pus and associated inflammation around the site of infection. It is particularly painful because of the pressure that builds up as infected material is trapped by the nail. Herpes simplex virus can also cause whitlows, although much less commonly than *Staphylococcus aureus*.

Staphylococcus aureus may cause more serious skin infections. Certain strains of this bacterium elaborate a toxin that causes inflammation of the skin, leading to a sloughing of the epidermis. Over large areas of the body, skin will peel off over the course of a couple of days. The associated reddening of the underlying tissue gives the condition its common name: scalded skin syndrome. Healthy tissue does regenerate and although this is not generally a life-threatening infection, babies who suffer from it may die through fluid loss.

A much more serious exfoliative condition caused by *Staphylococcus aureus* is toxic shock syndrome. This was first reported in the late 1970s. When it was first described, this very rare disease was seen only in menstruating women. It is particularly associated with the use of absorbent tampons containing rayon. Women suffering from this condition had often left a tampon in place for a considerable time, contributing to the risk of becoming ill. Toxic shock syndrome is associated with a high risk of mortality but it is no longer confined to menstruating women. Cases of men affected by this condition have been reported. It is caused by growth of *Staphylococcus aureus* strains that produce the toxin TSST-1. The toxin enters the bloodstream and causes symptoms of shock. Typically a patient suffers a sudden rise in body temperature accompanied by a skin rash that looks like sunburn. The patient suffers vomiting and diarrhoea and muscle pain as well as becoming dizzy. If treated in time, the patient's skin may peel off, particularly from the hands and feet. Treatment involves use of antistaphylococcal antibiotics. Corticosteroids may be used to relieve symptoms caused by the toxin but as this is a rare condition, the efficacy of this treatment has not been established in proper clinical trials.

Both *Staphylococcus aureus* and *Streptococcus pyogenes*, either alone or together, can cause **impetigo** although *Streptococcus pyogenes* is the most common cause of this infection. It is a relatively superficial infection affecting children and often seen on the face. Although it can develop very rapidly it does not often cause scarring of the skin. This may occur if the patient picks at the lesions, causing an opportunity for secondary infection with other pathogens. Impetigo caused by *Streptococcus pyogenes* is not associated with streptococcal throat infections. The strain isolated from a patient's skin lesion may, however, be found in the throat as well. Infection passes between children by direct contact with skin lesions. As with streptococcal throat infections, patients with streptococcal skin infections may develop glomerulonephritis as an immunological complication of the bacterial infection. Unlike streptococcal throat infections, the strains of *Streptococcus pyogenes* that infect skin are not likely to be associated with the development of rheumatic fever.

Streptococcus pyogenes produces virulence factors including hyaluronidase and streptokinase. These help break down connective tissues, thus aiding the spread of infection through the skin. **Cellulitis** is a more serious streptococcal skin infection. This is seen as a rapidly growing area of inflammation. Affected skin is very red and is hot to the touch. The underlying tissues become swollen and the lesion spreads with an alarming speed. If cellulitis is left untreated, the underlying tissues become infected and the patient develops streptococcal septicaemia. This carries a grave risk of mortality. Other bacteria including *Staphylococcus aureus*, *Haemophilus influenzae* and obligate anaerobes can occasionally cause cellulitis. Fournier's gangrene is an anaerobic cellulitis of the groin and scrotum. In immunocompromised individuals, *Pseudomonas aeruginosa* can cause a gangrenous lesion with marked tissue necrosis.

When streptococcal cellulitis affects the face it may be referred to as **erysipelas**. At first the rash may appear like sunburn spreading over the cheeks like a pair of butterfly wings, but as the disease develops the area becomes vesiculated and pustular. Erysipelas has also been referred to as St Anthony's fire, because of the burning sensation in the affected tissue. This is a confusing nomenclature since this name has also been used to describe ergotism.

An even more dramatic streptococcal infection is **necrotising fasciitis**. Although the initial infection is generally caused by *Streptococcus pyogenes* on its own, as the disease develops other bacteria may colonise and act synergistically to destroy tissues. Necrotising fasciitis starts as a rapidly swelling cellulitis but as fluid accumulates in the area beneath the skin the blood flow is impaired. This then provides an ideal environment for colonisation with anaerobic bacteria that destroy the sub-cutaneous tissues. The skin then separates from the underlying musculature. This is an aggressive infection in which the patient rapidly develops a potentially fatal septicaemia. If limbs are affected, radical amputation may help to prevent the rapid and unrelenting progress of this disease. Patients who do recover are left severely scarred. This has led to sensationalist headlines in tabloid newspapers such as '*FLESH EATING BUG ATE MY FACE*'.

Several clostridial species can cause an infection of deeper tissues known as gangrene. The most common cause is *Clostridium perfringens* but other species including *Clostridium septicum*, *Clostridium oedematiens* and *Clostridium histolyticum* can occasionally be isolated from gangrenous tissues. Clostridial spores enter traumatised skin and in people with poor blood circulation the relatively anaerobic environment in the tissues allows these bacteria to flourish. Although it is difficult to get *Clostridium perfringens* to spore in artificial culture its spores are widely distributed in Nature and can be found in the human faecal flora. Consequently, they can commonly be isolated from skin on the lower limbs.

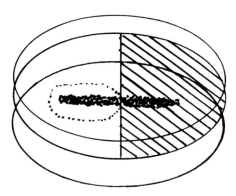

Fig. 7.5. The Nagler test. Half an egg-yolk plate is covered in antitoxin. The test strain is then inoculated from the untreated to the treated section of the plate. After incubation, strains of *Clostridium perfringens* can be identified because they produce a toxin that breaks down lecithin in the egg yolk to produce a region of precipitation that surrounds the microbial growth. This will only occur on the side of the plate that does not contain antitoxin. Antibodies in the antitoxin will bind to the toxin, preventing it from acting on the lecithin in the plate. This gives rise to the lollipop appearance shown here.

Clostridium perfringens can elaborate a number of toxins, but most tissue damage is caused by the α-toxin, a phospholipase. The laboratory test for *Clostridium perfringens* is based on the activity of this phospholipase (Fig. 7.5). This enzyme breaks down cell membranes causing extensive tissue damage. The infected area then becomes necrotic. This further reduces the blood supply, depriving the tissue of oxygen. In turn, this encourages growth of the bacterium and further cycle of damage and considerable pain for the victim. Gangrene caused by *Clostridium perfringens* is also associated with the appearance of gas bubbles in the infected tissue, caused by rapid bacterial metabolism using the muscle tissue as a substrate. This is why clostridial gangrene is also known as gas gangrene. Gangrene progresses very rapidly and is best treated by radical surgery. If a limb is affected then amputation is usually required and it is necessary to remove apparently healthy tissue to prevent spread of the infection post-operatively. If untreated, gangrene will lead to a toxic state and death.

Most of the skin infections considered so far may be characterised by the dramatic speed with which they can develop and spread. This is not always the case. Leprosy is a chronic skin infection caused by the intracellular pathogen *Mycobacterium leprae*. This bacterium cannot be grown in artificial culture but can be maintained in the footpad of immunosuppressed mice. It can also be grown in the nine-banded armadillo. This apparently strange choice of animal reflects the fact that *Mycobacterium leprae* has an optimal growth temperature lower than the human core body temperature. The nine-banded armadillo is the mammal with the lowest body temperature. Leprosy has been known since biblical times and has struck fear into generations of people. The ghost of Hamlet's father told the following of his death '...And in the porches of my ears did pour The leperous distilment...'. The very name 'leper' has long been a term of abuse implying someone who must be outcast from decent society. It is not, however, a particularly contagious disease and is often only acquired after prolonged close contact with an infected individual. It is a disease of overcrowding and poor hygiene.

After incubation for several years, leprosy causes a spectrum of disease manifestations that depend on the relative strength of the victim's cellular and humoral defences. Individuals who mount a strong cellular response to infection suffer from **tuberculous leprosy**. Infection causes discrete depigmentation of the infected skin and damage to the nerves in the underlying tissue causes a loss of sensation. **Lepromatous leprosy** causes extensive tissue damage with many organisms in infected areas. The skin becomes nodular and thickened with enlargement of the nostrils and cheeks. Infection causes insidious destruction of the nasal septum, causing the nose to collapse. Patients with lepromatous leprosy are said to have faces that resemble those of lions. Patients with leprosy often suffer damage such as burns. This is because infection with *Mycobacterium leprae* causes nerve damage and this is associated with an anaesthetic effect. Patients feel no pain, even when major trauma is inflicted. Secondary damage to leprous tissues can be very severe. Patients who have symptoms intermediate between tuberculous and lepromatous leprosy are said to have borderline leprosy. This may progress either to lepromatous or tuberculous leprosy depending upon the relative strength of the cell-mediated immunity of the patient. Since *Mycobacterium leprae* cannot be grown in artificial culture, laboratory diagnosis relies upon the demonstration of acid alcohol-fast bacilli in biopsy material or nasal scrapings. In tuberculous leprosy, bacteria may be very difficult to find but in lepromatous leprosy many bacteria are seen within infected cells. Leprosy is associated with characteristic tissue changes and histological examination of lesion material may confirm a diagnosis without a positive Ziehl Neelsen stain for acid alcohol-fast bacilli.

Other mycobacteria can cause skin lesions. *Mycobacterium tuberculosis* can very rarely cause a chronic ulceration of the skin, and scrofula is the condition caused when infection from a lymph node spreads to the skin. *Mycobacterium ulcerans*, as its name implies, causes a chronic ulceration of the skin. This bacterium and *Mycobacterium marinum* are both associated with skin infections in people who are frequently exposed to infected water. They cause conditions such as 'fish-tank granuloma' or 'swimming pool granuloma' depending upon the nature of exposure to the infection. These infections are confined to superficial tissues, reflecting their low optimum growth temperature.

7.8.2 What viruses cause skin lesions?

There are several virus infections that manifest in skin lesions. Some cause infections in the skin, others are more generalised infections that cause skin rashes as part of the disease process. Warts are benign and often pigmented tumours that crop on various parts of the body. These are caused by the papilloma family of viruses. Different members of the family cause warts in different anatomical sites. They are commonly found on hands and the feet. They may also be found around the anus and on the external genitalia. Genital warts carry an increased risk of developing cervical cancer in infected women. Warts on the feet are referred to as verrucas and can be very painful. Although warts appear and regress spontaneously, they have been subject to a wide range of folk remedies. These have included unspeakable acts involving toads to the application of salves made with vinegar and celendine extracts. Removal of verrucas even today is hardly a scientific process. They are commonly burned out, often using applications of liquid nitrogen to the wart and its surrounding skin. Care must be taken to ensure adequate treatment of the encircling tissues to make certain that the verrucas do not recur.

The orf virus is a pox virus that causes a pustular lesion on the hands of people who come into contact with sheep or goats. The pustules can become ulcerated. A single lesion generally develops on the hand but it may occur on an exposed forearm. The virus also causes infection in sheep. There is now a successful vaccine that controls orf infection in these animals. Since its introduction, orf virus has become a rare cause of human infection. Another pox virus causes molluscum contagiosum. This causes fleshy lesions up to 4 millimetres in diameter. This is an exclusively human disease and is spread by direct contact. It is seen most commonly among children. The lesions spontaneously regress within a few months.

Enterovirus infections are sometimes associated with skin rashes or reddening of the inside of the mouth. Of particular note, however, is hand, foot and mouth disease. This infection is caused by strains of coxsackievirus A. It causes a vesicular infection of the oral cavity. Vesicles are also found over the skin but concentrated particularly on the hands, feet and buttocks. Presumably modesty forbade mention of the last site when naming this disease.

Chicken pox is caused by the varicella zoster virus. This is a herpesvirus that can cause latent infection within the central nervous system. Primary infection usually occurs in childhood and upon initial infection the virus causes a crop of vesicles over the skin and mucous membranes. The fluid in each blister is laden with infectious virus. The skin lesions are very itchy and become pustular. Children find it very hard not to pick at the spots of chicken pox and almost everyone carries at least a couple of scars to remind them of their primary infection. These are the outward and visible signs of chicken pox. The virus, however, does not go away when the spots disappear. It travels up the peripheral nerves and into the central nervous system where it establishes an asymptomatic, latent infection. We develop immunity to varicella zoster virus, preventing a second attack of varicella, the posh name for chicken pox. Some considerable time after the primary infection, the virus may reactivate. Virus travels down a sensory nerve and causes a crop of vesicles over the skin served by that nerve. This is the zoster rash, commonly known as shingles. It affects a sharply defined area of skin: that served by the nerve down which it has travelled. The vesicles contain infectious virus. These will cause chicken pox in a susceptible individual. They do *not* cause shingles upon primary infection. In immunocompromised individuals, reactivation of shingles often involves more than one nerve and may be widely disseminated. Shingles is a very painful condition and the pain, a form of neuralgia, may persist for months after the skin lesions have healed. It is this neuralgia that is the most debilitating aspect of shingles and has led to a high proportion of suicides as a consequence of the pain.

Besides causing cold sores, herpes simplex virus type 1 can cause a severe infection in young children, spreading over large areas of skin. Indeed, the name of the virus derives from the Greek word *herpo*: to creep. There are eight human herpesviruses: herpes simplex virus types 1 and 2, varicella zoster virus, Epstein–Barr virus, cytomegalovirus and human herpesviruses 6, 7 and 8. Infection with human herpes 6 and 7 during childhood is associated with a transient skin rash that resolves, leaving a latent virus infection. As yet, scientists are uncertain as to the role of human herpesvirus 8, but it may play a role in the pathogenesis of Kaposi's sarcoma.

'Slapped cheek syndrome' is so called because of the bright red rash. This covers the face of the child suffering from this infection. It looks as if the victim has been slapped across the face. It is also known as 'fifth disease', being the fifth of six common rash-associated illnesses recognised by Victorian doctors. It is caused by a parvovirus: a very small single-stranded DNA virus.

Measles virus causes an infection of the respiratory tract. From there it spreads to cause a generalised infection associated with a characteristic skin rash. This starts to appear behind the ears and across the forehead, spreading over the face and then over the entire body and lasting for about five days. In well-nourished children, the disease is generally mild but in poorly fed individuals, measles is a much more aggressive infection and is associated with a significant risk of mortality. It has been estimated that more that 1.5 million children die each year from a measles infection. It affects the middle ear and it also causes conjunctivitis. In more serious cases this can lead to blindness. Patients with measles have minute white lesions covering the mucous membranes inside the mouth. These are referred to as Koplick's spots and are diagnostic for measles. Rubella is another generalised virus infection characterised by a skin rash, similar to that seen in measles. Most cases are very mild and the infection may be inapparent, revealed only after serological testing. The importance of rubella is that infection during pregnancy can lead to a devastating congenital infection. There is now a vaccine that protects against measles mumps and rubella. This exploits live, attenuated viruses.

7.8.3   What causes eye infections?

Chlamydia trachomatis is the cause of trachoma: the single most important eye infection affecting over 600 million people across the world and responsible for blindness in up to 20 million people. This is more than caused by any other single factor. Infection typically starts with the upper eyelid. This causes inflammation of the skin of the eyelid and of the conjunctiva of the eye. Rubbing of a sore eye transfers infection from one eye to another and a cycle of repeated infection is established. Tissues become scarred and prone to secondary bacterial infection. Different serovars of chlamydia can cause a less serious eye infection known as trachoma inclusion conjunctivitis, or TRIC for short.

People who wear contact lenses and who do not observe scrupulous hygiene run the risk of developing eye infections. Leaving contact lenses in place for too long can pre-dispose to a variety of infections as the protective tears fail to penetrate beneath the lenses. A serious infection that leads to severe corneal scarring is associated with *Acanthamoeba castellanii*. These are free-living protozoa that can infect lens wash fluid. If wash fluid is not changed regularly, then it may become contaminated and provide a potential source of eye infection.

Staphylococcus aureus, *Streptococcus pneumoniae* and *Neisseria gonorrhoeae* can all cause eye infections, particularly in the newborn. *Staphylococcus aureus* and

Streptococcus pneumoniae cause sticky eyes whereas *Neisseria gonorrhoeae* causes ophthalmia neonatorum. Besides causing eye infections, *Staphylococcus aureus* can infect the eyelid to cause 'styes'. *Pseudomonas aeruginosa* has been associated with serious eye infections caused by contaminated eye drops following eye operations. *Haemophilus influenzae* occasionally causes an aggressive eye infection. Conjunctivitis is a symptom of both secondary and congenital syphilis. Several viruses including adenoviruses, measles virus, enteroviruses, herpes simplex virus and varicella zoster virus may cause eye infections.

7.8.4   What animal-associated pathogens cause soft tissue infections?

A number of viruses associated with animals can cause serious diseases when they infect humans. These are often seen causing problems in exotic locations, and spread of these viruses typically involves an insect vector. Yellow fever was an infection confined to Africa but was introduced into the Americas through the slave trade. The first American cases were reported in the mid seventeenth century and hence it was an early importation. It is transmitted by mosquitoes (*Aedes aegypti*). In its most severe form, it causes a devastating liver disease with necrosis and associated jaundice, hence the name yellow fever. Liver damage leads to a loss of clotting factors in the blood and this is reflected by the propensity patients have for potentially fatal haemorrhage. Yellow fever induces a long-standing immunity in patients who survive. The last epidemic of yellow fever in Virginia occurred in 1855. Five of the six people infected in this episode who were tested for antibodies were found still to have circulating antibodies 75 years later. 'Jungle fever', caused by dengue virus, is closely related to yellow fever and is spread in the same way. Dengue haemorrhagic fever shock is a complication of infection with this virus and it carries a 10% risk of death. As there are four serovars of this virus, individuals in endemic areas may suffer repeated attacks. A number of other haemorrhagic viruses occur sporadically. These are often named after the geographical location in which they were first isolated or after the insect vector responsible for spread of the infection.

Marburg disease and Ebola fever are both caused by filamentous viruses found in Central an East Africa, although Marburg disease was first described causing infection in German laboratory workers exposed to African Green Monkeys. Both infections are severe with patients developing fever and profuse haemorrhage followed by disseminated intravascular coagulation. African Green Monkeys are not the natural host for these viruses and the ultimate reservoir has not been identified. It may well turn out to be a small rodent, as is the case with Lassa fever virus, endemic in parts of West Africa. This typically infects the bush rat (*Mastomys natalensis*). People who come into direct contact with the rat or its urine may develop a fever. This infection has a range of clinical symptoms. Patients may suffer a mild fever or progress through a spectrum of symptoms to severe illness associated with extensive haemorrhage, capillary damage and death.

Haemorrhagic infections are not simply confined to viruses. Rickettsias are small obligately intracellular bacteria that rely on the host cell to provide certain of their cellular functions. Apart from *Coxiella burnetii,* the cause of Q fever, all human rickettsial infections are transmitted by ticks, fleas, mites or lice and, with the exception of *Rickettsia prowazekii*, the cause of epidemic typhus, the insect vector is unaffected by the presence of rickettsias. The body louse, *Pediculus corporis*, is the vector of epidemic typhus and it dies within weeks of infection with the rickettsia.

Symptoms of typhus include prostration accompanied by a skin rash that may become haemorrhagic. Rickettsial diseases are sometimes known as spotted fevers for this reason and typhoid derives its name from the similarity of its rash to that of typhus. The two disease processes are, however, entirely distinct. Rickettsias are difficult to study and workers in the area, including Ricketts himself, have made the ultimate sacrifice for their research activities, dying to find out more about their subject.

Rickettsial diseases are infections associated with overcrowding and insanitary conditions and are particularly prevalent at times of war. Over one million soldiers were affected by trench fever during the First World War. This is caused by *Rickettsia quintana* and is spread by the body louse, *Pediculus corporis*.

As well as *Rickettsia prowazekii*, the cause of epidemic typhus, there are other species that cause different types of typhus. These include *Rickettsia typhi*, the flea-borne cause of endemic typhus; *Rickettsia tsutsugamushi*, the cause of scrub typhus and spread by mites; and *Rickettsia mooseri* spread by rat fleas and causing murine typhus. *Rickettsia rickettsia* is the cause of the tick-borne Rocky Mountain spotted fever. *Rickettsia conori* causes Mediterranean spotted fever, also known as African tick typhus. Rickettsial pox, another spotted fever, is caused by *Rickettsia akari*. Rickettsial infections may occasionally be relatively mild, as with rickettsial pox, but most are much more likely to be associated with a significant mortality.

Ticks and lice are responsible for spreading the spirochaetes responsible for relapsing fever. There are two forms of this disease. European relapsing fever is caused by *Borrelia recurrentis* and is spread by body lice (*Pediculus corporis*). Infection occurs when the louse is squashed, oozing its body contents over the skin of the victim. African relapsing fever is caused by *Borrelia duttoni* and this is spread by ticks from an animal reservoir. Armadillos are thought to act as a pool for tick-borne human infection. The main symptoms of relapsing fever are headache and fever with enlargement of the spleen. In patients who survive the first attack, the fever relapses after respite for about a week. This is thought to be caused by the emergence of bacteria with a variant antigenic structure.

Another infection that causes periods of fever alternating with periods of normal temperature is undulant fever, also known as brucellosis. This is caused by Gram-negative bacteria of the genus *Brucella*. Three species can cause human disease. *Brucella abortus* is acquired from cattle, *Brucella melitensis* is found in sheep and goats and *Brucella suis* comes from pigs. Farmers, veterinary surgeons and abattoir workers are most likely to contract brucellosis through contact with infected animals. Infections can occur through skin abrasions or through the alimentary canal. This disease was more common when the practice of drinking unpasteurised milk was more widespread. Most strains of *Brucella melitensis* and some strains of *Brucella abortus* cause an acute infection characterised by profuse sweating, fever and joint pains. *Brucella abortus* is more likely to cause a less acute infection that persists for longer than *Brucella melitensis*. Chronic brucellosis can persist for months or years, with recurrent attacks of a flu-like illness and a profound malaise and depression. It is chronic brucellosis that gives the disease the name undulant fever because of the alterations in body temperature over time as the disease waxes and wanes. Long-term antimicrobial chemotherapy, typically treatment for over three months, is necessary to cure the chronic form of this disease.

Anthrax is caused by the Gram-positive sporing *Bacillus anthracis*. Its spores are widely distributed in soils and can be found contaminating animal products including hides, bone meal and fleeces. Its most common presentation is cutaneous anthrax, seen in people who are regularly exposed to animal products. A very rare manifestation is the generally fatal pulmonary infection from inhaling spores. One folk name for this type of anthrax is wool sorters' disease because it was seen in people who handled contaminated fleeces. In its cutaneous manifestation the site of anthrax infection develops as a 'malignant pustule'. This is somewhat of a misnomer as no pus is produced. Over several days the lesion ulcerates and spreads causing tissue necrosis. The centre of the lesion is a shiny black. *Anthrax* is the Greek word for coal. In about 10% of cases of cutaneous anthrax, patients develop a septicaemia. Bacteria continue to multiply in the bloodstream and such patients are likely to die.

Tularaemia is another human bacterial infection caused by bacteria that normally infect animals. It is caused by the facultative Gram-negative

Francisella tularensis. It commonly infects species of rabbit, squirrels and other rodents and can be found in dogs and cats and the various arthropod parasites that live with these animals. The bacterium causes infection by contact with infected animals or their carcasses and is typically seen in hunters. Skin at the site of infection becomes ulcerated and the disease causes fever, swollen glands and enlargement of the liver and spleen. It is not an uncommon infection in North America and Scandinavia but is not seen in the UK. A close relative of *Francisella tularensis* is *Pasteurella multocida*. This bacterium is a common commensal in the upper respiratory tract of many animals including cats and dogs. Bites from these animals can often be infected by a mixture of bacteria, with *Pasteurella multocida* as the dominant pathogen. Rats carry two bacteria that can cause fever in humans who have been bitten. These are *Spirillum minus*, seen mainly in the Far East, and *Streptobacillus moniliformis*, the cause of rat bite fever described from America.

Leptospirosis is caused by spirochaetes of the genus *Leptospira*. These bacteria have characteristically slender spiral cells that are hooked at both ends. They can survive for long periods in stagnant waters. The most common infection caused by leptospires is Weil's disease, associated with *Leptospira icterohaemorrhagiae*. Rats are the reservoir for this infection and sewer maintenance engineers and cavers are prone to this infection, although it is also seen in people who habitually use freshwater rivers and lakes for leisure activities. It is characterised by a fever, jaundice and a severe conjunctivitis. The infection can be so severe as to cause liver failure and death, but this is rare.

Lyme disease is named after the Connecticut town in which it was first described in 1975. It is caused by the spiral bacterium *Borrelia burgdorferi* and is an infection that typically affects both mice and deer. It is the *Iodex* ticks of deer that transmit the infection to humans. Lyme disease is commonly seen in areas of forested land, such as New England in the USA and the New Forest in England. It is also more commonly seen in summer because people are more likely to walk around bare legged, giving the deer tick an opportunity to feed easily on humans. After an incubation period of about a week the patient suffers fever, headache, muscle pain and pain from swollen glands. A rash develops at the site of infection and this slowly enlarges. As it grows, the centre of the rash returns to a normal skin colour, leaving an enlarging red ring at the margins. The disease is associated with a number of serious complications. *Borrelia burgdorferi* is a rare cause of meningitis and encephalitis and has been reported to cause infections of the heart and pericardium. It is also associated with a persistent arthritis that can last for years. This is thought to result from immune complexes forming within affected joints because of

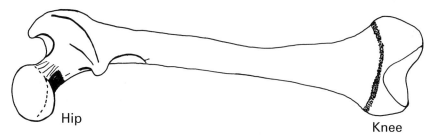

Fig. 7.6. Acute osteomyelitis commonly affects the growing points of long bones. The shaded region around the knee joint shows where a femur is commonly affected.

an antigenic cross-reaction between the host tissue at these sites and the bacterium.

7.8.5 What infections affect bone and joints?

Infections of the bone are referred to as osteomyelitis. This may either be an acute infection or a chronic disease. More than 90% of acute osteomyelitis cases are caused by *Staphylococcus aureus* but *Streptococcus pyogenes* and *Haemophilus influenzae* may also cause acute infection of the bone. Bone is affected because bacteria circulating in the bloodstream seed an infection. This is referred to as haematogenous spread of infection. Typically, acute osteomyelitis affects the growing points of long bones (Fig. 7.6). Blood flow in these regions is somewhat sluggish, allowing ample opportunity for the bacteria to settle and initiate infection at these sites. Although rare, osteomyelitis is most often seen in children and adolescents who are growing rapidly. Occasionally, acute osteomyelitis may be caused by spread of infection from adjacent tissues. In such cases, the infection is likely to be polymicrobial.

Chronic osteomyelitis is most often caused by *Mycobacterium tuberculosis* and is secondary to a pulmonary infection. The vertebrae are the most likely bones to be affected by chronic osteomyelitis but other bones including the hip, knee and bones of the hands and feet may be affected. Tuberculosis of the spine is most likely to be found in the thoracic region and the lesions can cause such pressure on the spinal cord as to cause paralysis below the affected region. Other bacteria causing chronic osteomyelitis include salmonellas and other coliform bacteria, *Pseudomonas aeruginosa* and the spirochaete *Treponema pallidum*. The last may cause bone lesions in children suffering from congenital syphilis. Alternatively, hypersensitivity may cause the formation of gumma on the bones of patients suffering from tertiary syphilis.

A number of pathogens can cause arthritis. The affected joint becomes swollen, painful and red. This may be a direct consequence of infection of the joint or as a result of an immunological reaction. In septic arthritis, joints become infected by bacteria that spread haematogenously or through direct inoculation through trauma. Prosthetic joints often fail because of chronic infections caused by coagulase-negative staphylococci introduced into the operation site when the artificial joint is put into place. The most common cause of septic arthritis in otherwise healthy joints is *Staphylococcus aureus*. Salmonellas and *Haemophilus influenzae* may cause septic arthritis in children. Septic arthritis is also a rare complication of gonorrhoea and is most often seen in women who otherwise have an asymptomatic *Neisseria gonorrhoeae* infection. Although septic arthritis typically affects one joint, a monoarthritis, in gonococcal arthritis several joints may be simultaneously be affected; this is polyarthritis. Arthritis may occasionally be caused by *Mycobacterium tuberculosis*. This chronic condition typically affects the hip or the knee joints.

Joints may become inflamed because of an immunological reaction to infection elsewhere in the body. A number of viruses cause reactive arthritis as a result of circulating immune complexes. Notable causes of reactive polyarthritis are hepatitis B virus and rubella virus. Polyarthritis caused by rubella virus can be seen following vaccination with the live, attenuated vaccine strain. Adult men who acquire a mumps infection may suffer a reactive polyarthritis along with other complications. Rheumatic fever is a rare complication of infection caused by *Streptococcus pyogenes*. This is an immunological disease affecting joints and endothelial tissues and caused by antigens that cross-react with streptococcal surface antigens and with host tissues. There is evidence to suggest that ankylosing spondylitis, a severe degenerative disease affecting the bones of the spine, may be associated with infections by a range of enteric organisms. These include species of the genera *Campylobacter*, *Klebsiella* and *Yersinia*. This condition is particularly seen in people with the HLA-B27 tissue type and is thought to be caused by cross-reacting antibodies in a manner somewhat similar to rheumatic fever. Another reactive arthritis is Reiter's syndrome, discussed in Section 7.3.4.

7.9 What causes perinatal infections?

The developing fetus poses an intriguing immunological problem. With half its genetic makeup coming from its father, a fetus represents 'foreign' tissue developing within the mother's womb. Normally, the mother's host defences would be activated to destroy tissue not recognised as 'self' but this does not happen to the fetus. Protective mechanisms have evolved that protect the fetus from its mother's defences. These have the bonus that they also protect the developing child from many infections. There are very few infections that affect the fetus but when an infection is established the consequences are generally catastrophic. One reason for this is the baby's phenomenal rate of growth. Also, the fetal tissues are very plastic and are in a constant state of flux as they develop. Infections, particularly those occurring early in pregnancy, may affect fetal tissue that would later become widely distributed through the body. Many fetal infections end either in a miscarriage or, when acquired later in pregnancy, in stillbirth.

If a woman is infected with rubella during the early stages of pregnancy, there is a very high risk that her infant will suffer severe symptoms of congenital rubella. As pregnancy proceeds, the risk of damage diminishes. It is in the first weeks that the fetus is at its most vulnerable. This is when fetal tissues are developing fastest and it is also a time when mothers may be unaware that they are pregnant. The clinical features of congenital rubella include deafness, blindness, heart defects and severe learning disabilities. Because of the serious nature of these defects pregnant women are monitored for rubella status throughout pregnancy. Women who are thought to have acquired rubella during the early stages of pregnancy may be offered a termination of pregnancy. Any women detected during screening who have not had rubella and who reach full term are offered vaccination after their baby is born. Vaccination for rubella involves a live, attenuated virus. It is not offered during pregnancy because of the possible risk to the developing fetus. At one time it was practice to offer vaccination to all young teenaged girls to prevent cases of congenital rubella. Now this vaccination programme has been extended to include all infants: boys as well as girls.

Although a significant proportion of the population has a latent cytomegalovirus infection, and despite its reactivation in a number of pregnant women, this virus does not generally cause significant problems for babies. There is some evidence that about 1% of babies excrete cytomegalovirus, probably following infection from a maternal reactivation. These babies may be slower to develop than unaffected children and are more likely to develop hearing difficulties but these can take a time to become apparent. Much more devastating are the congenital infections occurring in mothers who suffer a primary infection during pregnancy. In the worst cases, brain development is severely impaired and babies suffer jaundice, anaemia and spasticity. They also suffer from eye defects and have learning difficulties.

About one third of the population has had an asymptomatic infection with the protozoan *Toxoplasma gondii*. Occasionally this organism can cause an illness that is similar to glandular fever, with long-term profound listlessness. If a pregnant woman becomes infected with *Toxoplasma gondii*, however, the consequences can be much more serious. It is unusual amongst infections associated with pregnancy in that the most severe effects are seen when the mother is infected late in pregnancy. This is after much of the rapid fetal development is complete and the fetus is generally considered to be less vulnerable as a consequence. At birth, the baby may appear healthy but then develop hydrocephalus leading to learning disabilities. Children with congenital toxoplasmosis frequently suffer from eye defects as well.

Treponema pallidum can cause fetal infection, crossing the placenta relatively late in pregnancy. If a mother is treated for the disease in the early stages of pregnancy, congenital syphilis can be avoided. Consequently there is an active screening programme of pregnant women for syphilis. Congenital syphilis is now very rare in developed countries, but unfortunately it is more common in the Third World. Babies born with congenital syphilis have bone and tooth lesions, conjunctivitis and an enlarged liver and spleen. They often suffer from severe learning difficulties.

A mother infected with HIV has a 20% chance of infecting her infant with the virus, either *in utero* or shortly after birth. Such babies fail to thrive, have swollen lymph glands and an enlarged liver and spleen. They are prone to oral thrush. They generally develop full-blown AIDS within a year of birth. Another sexually transmissible infection, herpes simplex type 2 virus, does not cause congenital infection. If, however, a baby is born to a mother with genital herpes then the infant may become infected as it passes through the birth canal. It will then develop a severe, generalised fatal infection. This may be avoided if the baby is delivered by caesarean section rather than by a vaginal delivery.

It is not just developing babies who are susceptible to infection. During pregnancy reactivation of persistent or latent infections may occur, although they are generally asymptomatic. About 3% of pregnant women excrete cytomegalovirus in their urine as a result of reactivation of a latent infection. A similar proportion will excrete the polyomaviruses JC and BK during pregnancy. Women with genital herpes infections are more likely to suffer from active lesions when pregnant than otherwise and warts may be more common. During pregnancy, women are more vulnerable to infection and the course of the illness is generally more severe than for other people.

Women are also at risk from puerperal fever in the period immediately after birth. This is a life-threatening condition caused mostly by *Streptococcus pyogenes* but occasionally by other bacteria including coliforms, *Clostridium perfringens* and species of the genus *Bacteroides*. This condition has an important place in the history of infectious diseases. During the 1840s, Ignaz Semmelweis was puzzled by an apparent paradox. The wealthy women of Vienna who came into hospital to deliver their babies were far more likely to develop puerperal fever, also known as child-bed fever, than poor women in the same hospital. Women who were delivered by midwives in their own homes were also relatively safe from the disease. After long and detailed observation it occurred to Semmelweis that the rich patients were attended by medical students, a 'privilege' not extended to other women. He argued that these students were carrying 'cadaveric material' from the anatomy dissecting rooms and that this was the source of puerperal sepsis. He proved his theory by insisting that personnel should wash their hands in chlorinated lime before examining patients. This simple operation saved the lives of countless women and initiated one of the most important practices in the control of infection: hand washing.

7.10   What infections do fungi cause?

There is considerable variability in the ability of fungi to cause disease. Of the 250 000 or so fungi so far identified, only about 180 species are capable of causing disease in humans. Of these, very few are regularly isolated from infected human tissues. Most of the fungi that cause human infection are moulds, but there are pathogenic yeasts as well. Some pathogenic fungi are dimorphic. For some of the time they exist as yeasts but at others they adopt a mycelial form. When *Candida albicans* lives as a commensal it is typically seen as a yeast, but when it is actively causing disease, then it is much more likely to appear with mycelia. In contrast, many of the agents that cause systemic fungal infections are mycelial when free-living but then adopt a yeast form when causing infection.

Although the majority of fungi are harmless, some are highly pathogenic: capable of establishing a life-threatening infection in anyone who is exposed to the pathogen. These would include *Histoplasma capsulatum* and *Coccidioides immitis*. It was *Histoplasma capsulatum* that nearly claimed the life of Bob Dylan in 1997. Most fungal pathogens are, however, opportunists. They only cause infections in people who are in some way compromised. Fungal opportunist pathogens include species of the genera *Candida* and *Aspergillus*. The degree to which fungi can invade human tissue also varies. The name given to a fungal infection is a **mycosis** (derived from the Greek word *mukes* meaning a mushroom). Mycoses are often superficial. If they penetrate through the skin they are described as sub-cutaneous mycoses. These are more difficult to treat and can require surgical excision of the affected tissue to effect a cure. The most serious fungal infections are, however, the systemic mycoses.

Superficial fungal infections are very common. Dermatophyte fungi of several species can readily cause infection of skin, nails and hair causing a condition commonly and somewhat confusingly known as **ringworm**. This confusion persists in the Latinised medical names given to ringworm occurring on different parts of the body. *Tinea capitis* affects the head, *Tinea corporis* affects the body, *Tinea manuum* affects the hands, *Tinea cruris* affects the groin area and *Tinea pedis* affects the feet. *Tinea* is the Latin word for a worm or grub. It was originally thought that these infections were caused by worm-like parasites. Some of these infections have acquired descriptive common names. Athlete's foot is an infection that causes flaking and inflammation of the skin between the toes and is the most common fungal infection worldwide. Secondary bacterial colonisation of the lesions of athlete's foot may lead to weeping. Athlete's foot is easily spread when people share bathing facilities. Jock itch is the name given to fungal infections that affect the groin area. Several species of fungi cause infection of the skin, including those of the genera *Trichophyton*, *Microsporum* and *Epidermophyton*. Different species are found causing infection at different anatomical sites. Dermatophyte fungi typically cause discrete, scaly skin lesions that tend to be circular. It is from the shape of the lesion that the name *ring*worm is derived. Lesions tend to occur only on the torso or face; elsewhere ring-like lesions are rare. If ringworm affects the scalp it leads to patchy bald areas where the infection causes hair loss.

Dermatophyte fungi are moulds, but there are yeasts that can also cause skin infections. Pityriasis versicolor is a skin infection caused by the yeast *Malassezia furfur*. At one time this yeast was known as *Pityrosporum ovale* or *Pityrosporum orbiculare*. This depended upon the cellular morphology of the strain. Recent molecular studies have thrown the taxonomy of this pathogen into confusion. What is now recognised as *Malassezia furfur* is, in fact, a complex of seven species The significance of this latest taxonomic change has yet to be established. Normally, this yeast is a harmless skin commensal and the factors that cause it to become pathogenic are not understood. Infection causes either a bleaching of the skin or deposition of more skin pigment than usual together with flaking of the skin rash. Lesions are generally found on the trunk and they fluoresce a golden yellow colour when illuminated with ultraviolet light. *Candida albicans* does not usually tolerate the conditions provided by dry skin, but it can cause infections of wet skin. Lesions caused by *Candida albicans* are most often found where opposing skin surfaces provide a moist environment, especially in the armpit or under pendulous breasts. *Candida* species are best known for causing thrush, a common infection of the mouth or vagina where the fungus causes white patches on the mucosal surfaces.

There has been a recent dramatic rise in the number of life-threatening opportunist fungal infections. This has significantly increased the profile of fungi as pathogens. The rise in incidence is caused, at least in part, by the increase in immunocompromised individuals. A deficient immune system may result from infections. In people infected with HIV, the virus eventually damages immune defences. Malignancy and in particular leukaemia also results in damage to the functioning of the immune system. Damage may also be inflicted deliberately, for example through drug therapy to prevent organ rejection in transplant recipients. Such patients are at risk of developing generalised infections. It is immunocompromised patients who account for a rise in serious fungal infections. There has, however, also been a rise in the number of superficial infections such as ringworm and thrush. This increase is much more difficult to explain.

Systemic mycoses caused by primary pathogens are life-threatening deep-seated fungal infections. They may become widely distributed through the body and are generally initiated following the inhalation of the spores. A number of pathogenic dimorphic fungi are responsible for systemic mycoses. In endemic areas very many people have evidence of a primary lung infection. This is usually sub-clinical and the evidence that people have been infected comes from serological surveys. Only in a few people will the primary infection proceed to a secondary disseminated disease.

Systemic mycoses are mainly seen in the Americas although travellers to the Americas may export sporadic cases around the world. Coccidioidomycosis is caused by *Coccidioides immitis*. The principal symptoms include fever, cough, chest pains and weight loss. As the disease progresses, patients develop bone, joint and skin lesions. Chronic meningitis is not uncommon. Progress to death may be rapid or slow, with the disease course fluctuating. *Blastomyces dermatitidis* is the cause of blastomycosis. Skin lesions are the most common manifestation of this disease. These may vary from pimples to frank ulcers. These are associated with fever, weight loss and a productive cough. Patients may also suffer bone and joint lesions. The majority of *Histoplasma capsulatum* infections are sub-clinical or so mild that diagnosis may be difficult. Pulmonary histoplasmosis has a gradual onset with an increasingly productive cough, sometimes associated with night sweats. Disseminated histoplasmosis resembles miliary tuberculosis and may be associated with a chronic meningitis. Paracoccidioidomycosis, seen mostly in South America, is the consequence of infection with *Paracoccidioides brasiliensis*. Its symptoms include ulcers in and around the mouth and throat, enlarged lymph glands in the neck and skin lesions. A productive cough and difficulty in breathing associated with weight loss is a common feature of paracoccidioidomycosis. In fatal cases, lesions may be found in the adrenal glands and throughout the gastrointestinal tract.

Opportunist pathogenic fungi including species of the genera *Aspergillus, Candida* and *Cryptococcus* may cause systemic mycoses. These infections have a much wider geographical distribution and occur increasingly in patients who are compromised either by disease or drug treatment. Systemic opportunistic mycoses are among the most common causes of death in transplant recipients.

The successful diagnosis of fungal infections relies upon a combination of clinical observations and laboratory investigations. Clinical signs and symptoms can lead only to a presumed diagnosis of systemic fungal infections. Often the first indication that a patient has a systemic fungal infection is failure to respond to antibacterial chemotherapy. The earlier that a fungal infection is investigated, the greater is the chance of a successful outcome for the patient. This is especially important for patients at increased risk of developing systemic mycoses.

Candida species are now the most common cause of fungal infection in patients who are immunocompromised. *Candida albicans* is the most frequent cause of systemic candidosis but other species, including *Candida tropicalis* and *Candida glabrata*, are increasing in importance as the cause of opportunistic infection. Systemic candida infections may become localised, particularly if the patient is receiving corticosteroids to prevent transplant rejection or cytotoxic drugs to treat malignancies. A common site for localised deep-seated candida infection is within the kidney. Such infections generally arise as a result of haematogenous spread during systemic infection rather than by ascending from the bladder. Lesions may also develop in the liver, spleen, brain, peritoneal cavity and gastrointestinal tract. Endocarditis can follow candida infection of a heart valve. Candida infections may also be widely disseminated. Pulmonary infections caused by *Candida albicans* are rare. One of the common signs of deep-seated candida infection is the appearance of fluffy white lesions within the eye. These look like cotton wool and are referred to as candida endophthalmitis.

Deep-seated infections caused by *Candida albicans* are difficult to diagnose and treat. Because treatment is often delayed and since these infections are associated with patients who are already seriously ill, the **prognosis**, predicted outcome of the disease process, is generally poor in these cases. Many cases of deep-seated infection caused by *Candida albicans* only become apparent at *post mortem* examinations. Laboratory diagnosis of candidosis is complicated since *Candida* species form part of the human commensal flora at a number of sites. These may contaminate clinical specimens. Unless it is isolated from a site that is normally sterile, the presence of a single isolate of a *Candida* species is of little significance. Repeated isolation of the same yeast, preferably from different sample sites, would, however, support the diagnosis of candidosis. This may be further supported by a rise in the number of organisms isolated over time. Antibody tests are of limited value since individuals may mount an immune response to commensal yeasts and most immunocompromised people do not produce a detectable antibody response.

Systemic candidosis is treated with a combination of amphotericin B and flucytosine. Therapy with flucytosine alone is not used because of the problem of resistance emerging during therapy. Because of its solubility, fluconazole has been used successfully to treat urinary tract infections and peritonitis caused by *Candida albicans*. Itraconazole has also been used successfully in certain cases.

Severely immunocompromised patients with serious underlying disease, especially malignancy, are at risk of developing invasive aspergillosis. The chances of such an infection are significantly increased if the patient is receiving cytotoxic or immunosuppressive therapy for their underlying medical condition. Treatment with corticosteroids coupled with a low leukocyte count is the most important pre-disposing factor for the development of invasive aspergillosis. It has been estimated that about one third of patients who have had heart, liver or bone marrow transplants will have evidence of an invasive aspergillus infection at *post mortem* examination. A similar proportion is expected in patients with acute leukaemia.

Air-borne aspergillus spores are the source of infection and in about 70% of patients with invasive aspergillosis, the lung is the only organ affected. *Aspergillus fumigatus* is the species most frequently isolated from lesions. *Aspergillus* species may cause widespread destruction of lung tissue. This leads to an invasion of blood vessels where the infection causes **thrombosis**. Septic emboli may break off and spread the infection to other organs, notably the kidneys, heart and brain. The prognosis is poor for a patient with invasive aspergillosis and the condition is often only diagnosed at *post mortem* examination. The treatment of choice for invasive aspergillosis is intravenous amphotericin B. Itraconazole has been used to treat aspergillus infections but has not yet been fully evaluated in the treatment of serious systemic infection.

7.10.1   How are mycoses diagnosed in the laboratory?

Laboratory diagnosis of systemic mycoses depends upon the exploitation of a number of techniques. These include microscopic visualisation of the pathogen, its isolation in artificial culture and the use of a variety of serological tests. The weight given to the results of any individual result varies depending upon the type of infection and its causative organism. Close liaison between the laboratory and the clinician is essential in establishing the correct diagnosis.

It is important that the laboratory receives good quality specimens of the correct type and that these are accompanied by good, relevant clinical histories. Samples should be delivered as quickly as possible to the laboratory. This is to prevent overgrowth of organisms such as *Candida albicans* in specimens. These can grow rapidly in clinical material left at room temperature. Urine specimens are best collected through a newly inserted catheter or by suprapubic aspiration of the bladder. Bronchial secretions or bronchoalveolar lavage specimens obtained by endoscopy provide more reliable results than do sputum specimens. These are frequently contaminated by organisms from the throat and mouth. In difficult cases, lung biopsy samples may be required to establish the correct diagnosis.

Most pathogenic fungi grow well upon laboratory media such as Sabouraud glucose agar or 4% malt extract agar. When used to isolate fungi from clinical material, chloramphenicol is often added to the isolation medium to prevent the overgrowth of bacteria that may also be present in the sample. When trying to isolate a dermatophyte, cycloheximide may also be added to prevent the growth of saprophytic fungi. Many fungal pathogens have an optimal growth temperature below the normal human body temperature of 37°C. Consequently, fungal cultures are routinely incubated at 25–30°C as well as at 37°C. Many fungi grow relatively slowly when compared with bacterial growth, and fungal cultures are retained for at least two to three weeks. In some cases, cultures are kept for up to six weeks. Yeasts generally grow within five days and some may grow overnight.

Pathogenic moulds are identified by their colony morphology and their microscopic appearance. Yeast identification is more akin to the identification of bacteria and relies upon fermentation and assimilation tests, together with other biochemical features. Commercial kits are available for the identification of pathogenic yeasts. Rapid tests can be used to identify *Candida albicans*. This is the only species of the genus *Candida* to produce a **germ tube**. When *Candida albicans* is incubated in serum at 37°C for about two hours, the yeast cells begin to produce mycelia, referred to as germ tubes. Observation of germ tube production is useful since *Candida albicans* accounts for between 80 and 90% of yeasts isolated from clinical material.

The results of fungal cultures must be treated with caution. They may provide unequivocal evidence of infection when an established primary fungal pathogen or an opportunistic fungus is isolated from a site that should be sterile in health. There is much greater difficulty when isolating a commensal fungus with the capacity for causing opportunistic disease. It should also be remembered that a negative culture result does not prove absence of an infection.

An array of serological tests have been developed to augment fungal culture. These rely upon the detection either of fungal antigens or of antibodies raised against fungi. Techniques exploited to detect antibodies raised against fungi include immunodiffusion, counter-current immunoelectrophoresis, whole-cell agglutination, complement fixation and ELISA. Fungal antigens may be detected using latex particle agglutination, ELISA and radioimmunoassay. Certain tests lack sensitivity; others are plagued by problems of specificity. The value of serological tests varies and are best used in conjunction with other laboratory and clinical findings.

Molecular biological techniques are currently being developed to assist in the diagnosis of opportunist fungal infections. One important development is the application of PCR technology. The value of such tests is yet to be established. Because of the extreme sensitivity of this technology, care must be taken in interpreting the results obtained from specimens that may carry commensal fungi, even in very low numbers.

Having successfully established the diagnosis of a systemic fungal infection, laboratory personnel and the clinician who is managing the patient may further collaborate. The laboratory has a role in advising on appropriate chemotherapy. Furthermore, the laboratory may assist in monitoring treatment. Routine susceptibility testing is only required for flucytosine and fluconazole and itraconazole sensitivity. In patients receiving flucytosine it is also necessary to monitor the serum drug level to ensure that toxic levels do not accumulate. This is particularly important for patients who have renal impairment.

Candida species grow readily in cultures incubated at 37°C and can be isolated on common media such as Sabouraud agar. The most frequent marker of systemic candidosis is the presence of this fungus in a blood culture. It may be necessary to make repeated blood cultures. Up to half the cases of systemic candidosis proven at *post mortem* examination are blood culture negative. There is a further complication. Patients with indwelling intravenous catheters often suffer transient episodes of fungaemia, when *Candida* species may be isolated from the blood without organ involvement. Yeasts are often present in the urine of patients with systemic candidosis. Because of the commensal status of the genus *Candida*, however, their isolation can only be used to support a diagnosis, not to confirm it.

The laboratory diagnosis of aspergillosis is difficult, and invasive aspergillosis is particularly problematic. In cases of aspergilloma, sputum samples are best studied after digestion with pancreatin and treatment with potassium hydroxide. The fungus may be difficult to find but when it can be observed it is often present as solid wefts of mycelium that have broken free of the fungus ball. Mycelia are divided into sections by septa and they display a characteristic branching structure. Within sputum, sporing structures are very rarely seen. Patients with allergic aspergillosis usually have abundant fungus in their sputum samples. The only reliable method of diagnosing invasive aspergillosis is using a tissue biopsy sample, a procedure that carries a significant risk for the patient. Serological tests alone are unreliable.

Aspergilli can be easily isolated in artificial culture using Sabouraud agar without cycloheximide. Colonies will appear after one or two days when cultures are incubated at 25–37°C. Quantitative sputum cultures yield the best diagnostic information. Following digestion to liquefy the specimen, it is centrifuged and the sediment is resuspended in saline solution. Culture of serial dilutions allows an estimate of the number of fungi present in the original sample. Cultures taken before death from patients with invasive aspergillosis are positive in fewer than 35% of cases, even in the presence of widely disseminated disease. Patients with aspergilloma typically have low fungal counts while those with allergic aspergillosis yield large numbers of fungi.

7.11 How do we recognise clinically important bacteria?

For more than a century bacteria have been classified according to their 'Gram reaction', named after Christian Gram who developed the staining protocol in 1884. Cells are stained with a crystal violet–iodine complex and are then exposed to an organic solvent, usually acetone. Gram-positive bacteria retain the complex and appear blue-black or purple when viewed microscopically. In Gram-negative bacteria the complex is leeched out and they require counterstaining with a red dye before they are easily seen under the microscope. Another important feature used in the preliminary identification of bacteria is their shape. Most bacteria are either round (cocci, singular coccus), rod-shaped (bacilli, singular bacillus) or spiral. A great deal of information can be gathered just by observing the Gram reaction and the shape of a clinical isolate. Laboratory identification of clinically important bacteria is an important prerequisite for effective treatment of many infections. Figure 7.7 shows the steps involved for Gram-positive and Gram-negative bacteria of medical importance.

(a) Gram-positive bacteria

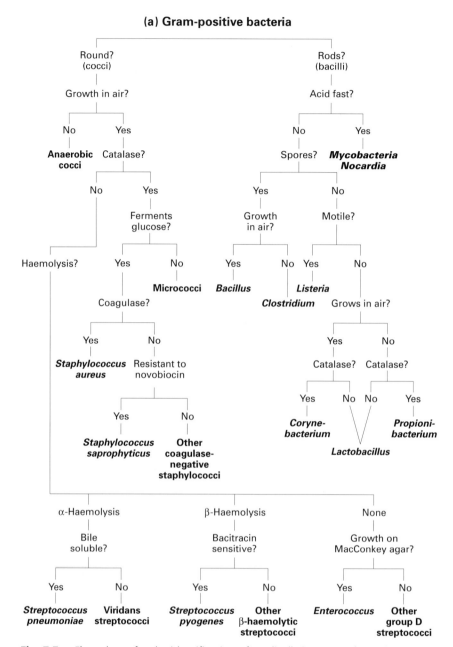

Fig. 7.7.   Flow charts for the identification of medically important bacteria: (a) Gram-positive bacteria; (b) Gram-negative bacteria.

(b) Gram-negative bacteria

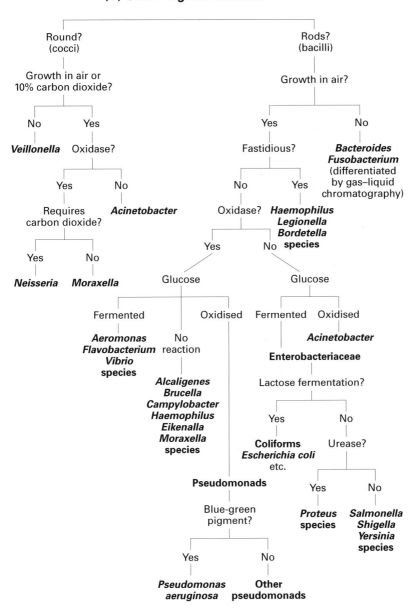

Fig. 7.7 (*cont.*)

The Gram stain involves making a light suspension of the sample to be tested in a drop of saline solution on a microscope slide. This is allowed to dry in the air and the specimen is then fixed to the slide by passing it three or four times through a hot Bunsen burner flame with the sample uppermost. Once the slide has cooled it is flooded with crystal violet and left for one minute. The crystal violet is washed off and a solution of Lugol's iodine is applied for another minute. Again the slide is washed and the excess water is drained. The critical step in Gram staining is the decolorisation. Acetone is applied and washed off immediately with plenty of water. If the acetone is not applied for long enough, Gram-negative bacteria will retain sufficient crystal violet–iodine complex to appear Gram positive. Conversely, if the acetone is left for too long then the dye complex may leech out of Gram-positive cells. To allow Gram-negative cells to be easily visualised a counterstain such as carbol fuchsin is applied for about 30 seconds. Once the film has been dried it is ready for examination.

7.11.1 Gram-positive cocci

There are two groups of Gram-positive bacteria that are important in medical microbiology: staphylococci and streptococci. *Staphule* is Greek for a bunch of grapes and *streptos* is another Greek word meaning twisted and implying a chain. Staphylococci do tend to form bunches and streptococci also tend to line up in chains. These microscopic cellular arrangements are more clearly seen when liquid cultures are used to make preparations. It can, however, be difficult for microbiologists in training to make good microscopic preparations from liquid cultures. Although their names imply a morphological difference, isolates are more accurately differentiated by the catalase test. Using a wide-bore capillary tube a small quantity of the culture to be tested is picked from the plate. The other end of the tube is inverted into hydrogen peroxide. This is then allowed to run down the tube and onto the culture material. Catalase-positive staphylococci will cause oxygen bubbles to be released from the hydrogen peroxide and these can clearly be seen rising rapidly through the tube. The catalase-negative streptococci cannot do this. Care must be taken with test since fresh blood contains catalase and streptococci tested from fresh blood agar may give a false-positive reaction.

Staphylococcus aureus is clinically the most important of the staphylococci. Its species name implies that colonies are a golden yellow colour (*aurum* is Latin for gold). Although this is true of many strains and even though this colour may be enhanced if cultures are grown on a lipid-rich medium like milk agar, not all clinical isolates show this classic characteristic. Coagulase-negative staphylococci grow as white colonies. Indeed, at one time, all the coagulase-negative staphylococci were grouped together and given the name *Staphylococcus albus*, the white staphylococcus.

Certain strains of *Staphylococcus aureus* causing serious disease can be indistinguishable from coagulase-negative staphylococci when simply observing colony morphology. The coagulase test is used to differentiate *Staphylococcus aureus* from other staphylococci. This is the enzyme produced by *Staphylococcus aureus* that causes plasma to clot. Many laboratories employ a slide test to screen for 'clumping factor' although this is not true coagulase. All slide test-positive isolates are regarded as *Staphylococcus aureus*. Three drops of saline solution are placed on a clean microscope slide. A strain of *Staphylococcus aureus* is suspended in the first drop to act as a positive control. The test strain is evenly suspended in the remaining two drops. Using a sterile loop, a drop of citrated rabbit plasma is added to the positive control and to one of the test suspensions and the drops are mixed thoroughly. Clumping should be seen within ten seconds if the test is positive. The untreated control is used to show that the test strain does not clump spontaneously. If this screen is negative then a tube coagulase method can be employed. Colonies to be tested are emulsified in a 1:10 dilution of citrated plasma in saline solution. The tubes are incubated at 37°C and are examined after one hour, six hours and after overnight incubation. A coagulase-positive strain will cause conversion of the plasma into a soft or stiff gel, seen on tilting the tube to a horizontal position.

A very few *Staphylococcus aureus* strains fail to produce coagulase, as shown by either the slide or the tube method. These may be identified by the DNAse test. Colonies to be tested are plated onto tryptose agar containing DNA together with a known *Staphylococcus aureus* to act as a positive control and a coagulase-negative staphylococcus used as a negative control. The culture is incubated overnight at 37°C and following incubation the plate is flooded with 1 molar hydrochloric acid. This causes precipitation of DNA within the agar, turning the medium cloudy. DNAse-positive *Staphylococcus aureus* isolates will be surrounded by a clear halo where the DNA has been broken down.

Streptococci are classified by their ability to cause the breakdown of blood in fresh blood agar plates. The non-haemolytic streptococci do not cause alteration in the medium. The most important of these are the enterococci, including *Enterococcus faecalis* and *Enterococcus faecium*. These have only recently been separated taxonomically from the streptococci as a result of molecular studies. These bacteria are, as the name implies, found in the gut and can grow in the presence of bile salts.

Streptococci that can cause partial breakdown of blood are known as the α-haemolytic streptococci. Colonies growing on fresh blood agar are surrounded by a greenish halo of partial haemolysis. There is one α-haemolytic streptococcus that must be differentiated from the others. This is *Streptococcus pneumoniae*: the remaining species are grouped together as the viridans streptococci. *Viridis* is the Latin word for green. *Streptococcus pneumoniae* is bile soluble and sensitive to optochin (ethylene hydrocupreine hydrochloride) whereas the viridans streptococci are not soluble in bile and are resistant to optochin, an antibiotic too toxic for therapeutic use.

Susceptibility to optochin is tested using a standard controlled disc diffusion test. A filter paper disc soaked in optochin is placed at the junction of a known *Streptococcus pneumoniae* strain and the isolate to be tested. After overnight incubation, a zone of clearing around the disc indicates *Streptococcus pneumoniae*. Viridans streptococci can grow up to the disc. The use of a previously characterised strain of *Streptococcus pneumoniae* gives confidence that the disc contained active antibiotic since this will not be able to grow up to the disc. If a pure culture of the test organism is available, the bile solubility test is performed by making a smooth suspension of the culture to be tested in a 10% deoxycholate solution held at 37°C. Clearing should be seen within 30 minutes if *Streptococcus pneumoniae* is tested. In a mixed culture, a drop of deoxycholate solution is applied to the suspect colony. This should lyse within 30 minutes if it is *Streptococcus pneumoniae*. This method is not entirely satisfactory since it is difficult to use controls and it is better to defer the test to work with pure cultures, if possible.

The β-haemolytic streptococci can break down fresh blood completely. When trying to decide which group of streptococci have the greater effect on blood remember that 'beta is completa'. Colonies of β-haemolytic streptococci growing on fresh blood agar are surrounded by a halo of complete clearing. The size of this zone may be increased if the culture is incubated anaerobically as one of the enzymes responsible for haemolysis, streptolysin O, is inactivated by oxygen. Clinically the most important of the β-haemolytic streptococci is *Streptococcus pyogenes*. This species is classified in the Lancefield group A based upon its antigenic structure. This can be tested using a latex agglutination method. Latex beads coated with antibodies raised against the various Lancefield group antigens are added to smooth suspensions of the bacteria to be grouped. If the antigen on the bacterial cells matches the antibody coating the latex, particles will clump together for that particular antigen–antibody combination. *Streptococcus pyogenes* is also sensitive to the antibiotic bacitracin whereas other β-haemolytic streptococci are resistant to this agent.

7.11.2   Gram-positive bacilli

An important characteristic used to separate the Gram-positive bacilli taxonomically is the ability to produce spores. These may be visualised microscopically using a modification of the Ziehl Neelsen stain, used to visualise mycobacteria. A fixed, light suspension of the culture to be examined is stained for three to five minutes with a strong carbol fuchsin solution. During this time the slide is heated so that steam rises from the stain. This allows the dye to enter any spores that may be present. Care must be taken not to boil the stain since this would destroy the preparation by denaturing the cells. The slide is then washed thoroughly in water and is treated for between 15 seconds and one minute with 0.25% sulphuric acid. This will allow decolorisation of most structures. The spores, however, will protect the carbol fuchsin from acid decolorisation. The acid is then washed off and the preparation is counterstained with methylene blue. Alternatively, malachite green may be used as a counterstain. The chosen counterstain is washed off and the slide dried and viewed microscopically. Spores are seen as red structures and other material stains blue or green according to the choice of counterstain. In the original Ziehl Neelsen method, a 3% acid alcohol mixture replaces the sulphuric acid used to decolorise the preparation in the spore stain modification. The waxy cell walls of mycobacteria can resist even this attempt to decolorise carbol fuchsin.

There are two groups of sporing Gram-positive bacilli: the clostridia and the somewhat confusingly named genus *Bacillus*. Clostridia are obligate anaerobes. They include some important human pathogens including *Clostridium perfringens*, causing gas gangrene; *Clostridium tetani*, the cause of tetanus; and *Clostridium botulinum*, the cause of botulism. At one time *Clostridium perfringens* was known as *Clostridium welchii* and references are still occasionally made to this nomenclature by older clinicians. Its spores are rarely seen in cultures. In contrast, the spores of *Clostridium tetani* frequently cause a marked swelling at the end of the bacterial cell, giving it the appearance of a drumstick. The Nagler test is used in the identification of *Clostridium perfringens* (Fig. 7.5). This relies upon the demonstration of the α-toxin of this bacterium. This toxin acts as a phospholipase that can break down lecithin in, for example, egg yolk. In the Nagler test antitoxin raised against the *Clostridium perfringens* α-toxin is spread over one half of an egg yolk plate and allowed to dry. The plate is then inoculated with the strain to be tested, together with appropriate controls. It is then incubated anaerobically. Isolates of *Clostridium perfringens* cause precipitation in the untreated side of the plate but not in the area that has had the application of antitoxin.

The aerobic sporing Gram-positive rods belong to the genus *Bacillus*. At one time this genus name was applied to many other rod-shaped bacteria: *Bacillus coli* and *Bacillus typhi*, for example. Taxonomic methods have been refined and these bacteria will now be much more familiar as the Gram-negative *Escherichia coli* and *Salmonella typhi*. Another confusion lies with the continued use of the name 'bacillus'. Does it refer to the genus, in which case its initial letter is capitalised and the word is italicised in print; or does it refer to any rod-shaped bacterium? If so, the word is not italicised and does not require a capital initial letter. In speech we are not given these valuable visual clues. The genus *Bacillus* has two species that are clinically important. *Bacillus anthracis* causes anthrax and *Bacillus cereus* causes food poisonings. It is ironic that this bacterium is named after Ceres, the Roman goddess with special responsibility for the harvest.

The non-sporing Gram-positive rods are often differentiated by their motility. A drop of liquid culture is placed on a microscope coverslip and this is inverted over a plasticine ring on a slide. Motility is best viewed using phase-contrast or dark-ground microscopy and care should be taken to distinguish true motility from the random Brownian motion that affects all small particles. When a bacterium is truly motile there will be an apparent movement in particular directions: Brownian motion involves oscillations around a fixed point.

Motility distinguishes the genus *Listeria* from the coryneforms and the lactobacilli. This distinction is not absolute, however. *Listeria monocytogenes* is an important human pathogen causing septicaemia and meningitis, particularly in the newborn and in people who are immunocompromised. At 25°C, this bacterium exhibits a characteristic tumbling motility. At 37°C it is not motile.

Lactobacilli are not generally considered pathogenic but they play an important role in protecting against infection. They are a major component of the vaginal flora of women in the reproductive years. They often appear as long, slender Gram-variable rods when viewed microscopically. They tend to make their local microenvironment too acid for other bacteria to tolerate. Lactobacilli are catalase-negative; this feature is used to differentiate them from the catalase-positive non-motile coryneform bacteria. The most infamous coryneform is *Corynebacterium diphtheriae*, toxigenic strains of which cause diphtheria. Consequently, coryneform bacteria are also known as diphtheroids. They tend to have irregular cellular shapes and cluster together in a characteristic manner, described by some microbiologists as Chinese letters. Propionibacteria are coryneforms that cannot grow in an oxygen atmosphere. A conspicuous example is *Propionibacterium acnes*: the bacterium notably associated with acne.

7.11.3   Mycobacteria

Through a consideration of the structure of the cell wall, mycobacteria are classified together with the Gram-positive bacteria, but the waxy nature of the mycobacterial cell wall prevents them from staining using conventional Gram-staining methods. To be easily visualised they require the extremes of the Ziehl Neelsen staining protocol described above. Mycobacteria are sometimes referred to as acid alcohol-fast bacteria because of their resistance to decolorisation in this method. Important pathogens include *Mycobacterium tuberculosis*, the cause of tuberculosis, and *Mycobacterium leprae*, the cause of leprosy; with the increase in HIV infection, bacteria of the *Mycobacterium avium-intracellulare* complex are rapidly increasing in importance as opportunist pathogens. Mycobacteria are very slow growing, some taking several weeks to yield visible colonies on laboratory media and *Mycobacterium leprae* cannot grow at all on artificial media.

7.11.4   Gram-negative cocci

The most clinically important Gram-negative cocci all belong to the genus *Neisseria*. They are *Neisseria gonorrhoeae*, the cause of gonorrhoea, and *Neisseria meningitidis*, the causative agent of meningococcal meningitis. Members of this genus tend to form pairs of cells that look rather like kidney beans and they are often referred to as diplococci. They are very vulnerable to drying. In the laboratory they must be grown under a humid atmosphere in which the carbon dioxide concentration is raised to between 5 and 10%.

7.11.5   Gram-negative bacilli

The Enterobacteriaceae are a large family that contains several medically very important bacteria. They are described as facultative anaerobes since they can grow in the presence or absence of oxygen. As the name implies, the natural habitat for these bacteria is in animal guts. Despite this, another group of Gram-negative bacilli, strict anaerobes such as those of the genus *Bacteroides*, are numerically far more common in the bowel flora than are the facultatively anaerobic Enterobacteriaceae. Notable examples of bacteria belonging to the family Enterobacteriaceae include *Yersinia pestis*, the cause of plague and enteric pathogens including salmonellas and shigellas. This family also includes *Escherichia coli*: arguably the most extensively studied microbe on the planet.

Members of the Enterobacteriaceae are differentiated largely by variations in their metabolic profiles and their antigenic structures. There are more than 1800 separate antigenic variants of salmonella, for example. Used in the identification of Enterobacteriaceae to genus level, one important metabolic difference is the ability or otherwise to ferment lactose. This test is used to separate faecal pathogens such as salmonellas and shigellas, unable to ferment lactose, from commensal gut bacteria such as *Escherichia coli* and the klebsiellas that can ferment lactose. This test is so important that MacConkey's agar and broth were developed to help to isolate and identify enteric pathogens. MacConkey's medium contains bile salts to inhibit the growth of non-enteric bacteria. Peptone provides a nutrient source for bacteria but these media also contain lactose and neutral red, a pH indicator. If a bacterium can ferment lactose, enough acid is produced to drop the pH to a point where colonies growing on MacConkey's agar take on a red colour and MacConkey's broth turns from purple to yellow. Non-lactose fermenters cannot do this and they rely on the metabolism of peptone to grow. This raises the pH in the medium and the colonies on a MacConkey plate appear a buffish yellow colour.

Members of the Enterobacteriaceae can also be identified by other characteristics. Members of the genus *Proteus* are so highly motile that, following overnight incubation, a single colony can swarm over the entire surface of an agar plate. This genus takes its name from Proteus, the old man of the sea in Greek mythology. He escaped his enemies through his miraculous ability to change shape. Members of the genus *Proteus* adopt the shape of the culture dish, whatever that may be. Few bacteria produce striking pigments. Another of the Enterobacteriaceae, *Serratia marcescens*, provides a conspicuous exception. This produces prodigiosin, a blood-red pigment, particularly when it is growing at temperatures of about 25–30°C. It has been suggested that growth of *Serratia marcescens* on the Eucharistic bread may account for at least some of the claims for mediaeval transubstantiation miracles: the conversion of bread into flesh. Its growth does resemble drops of blood when it is deliberately inoculated onto bread. One such miracle has been immortalised in a fresco in the Vatican: 'the Mass of Bolsena' by Raphael.

The Enterobacteriaceae do not elaborate the enzyme complex known as oxidase. This is cytochrome oxidase, an enzyme used in respiration. Pseudomonads are obligate aerobes that are oxidase positive. Colonies of pseudomonads can be difficult to differentiate from those of some

Enterobacteriaceae, but the oxidase test may be used to distinguish these bacteria. A drop of freshly prepared 1% tetramethyl-*para*-phenylene diamine dihydrochloride (TMPPD, or oxidase reagent for short) is used to wet a sterile cotton wool swab. Part of the freshly grown colony to be tested is applied to the wetted swab. If the bacterium is oxidase positive, a deep purple colour will appear within ten seconds. This test was performed by rubbing a colony onto a wetted filter paper strip using a bacteriological loop. However, the nickel used to make many bacteriological loops may react with oxidase reagent giving a false-positive result if a wire loop is used in the oxidase test. Clinically the most important of the pseudomonads is *Pseudomonas aeruginosa*, an opportunist pathogen that causes wound infections that are difficult to treat. This species produces pyocyanin, a soluble blue-green pigment that aids its identification.

Some Gram-negative rods are so short that they look like cocci when viewed using a light microscope. These are sometimes referred to as coccobacilli as a result. An important example is the genus *Acinetobacter*. These bacteria are increasingly being recognised as the cause of hospital-acquired infection. They can cause a wide range of infections, particularly in patients needing intensive-therapy facilities.

Some Gram-negative bacilli are very fastidious in their growth requirements. They can only be cultivated artificially if all their nutritional demands are met. This, however, may be exploited in their identification. Members of the genus *Haemophilus* have strict nutritional requirements. *Haemophilus influenzae* requires two growth factors, X and V, to grow. The X factor has now been identified as haem and the V factor is nicotinamide adenine dinucleotide (NAD). In contrast to the pathogenic *Haemophilus influenzae*, its commensal cousin *Haemophilus parainfluenzae* requires only the V factor. It can make its own haem.

The fastidious nature of some Gram-negative bacteria can cause severe problems for their artificial culture. Although members of the genus *Legionella* can be provided with sufficient iron and cysteine to allow them to grow, they rapidly produce toxic metabolic products that inhibit their growth. Colonies will only grow well on media that contain charcoal. This adsorbs the toxic compounds permitting artificial culture of these bacteria. A similar problem afflicts *Bordetella pertussis*, another fastidious Gram-negative bacillus.

Not all Gram-negative bacilli are straight. The vibrios are a collection of curved Gram-negative rods. The archetypal example of a vibrio is *Vibrio cholerae*, the pathogen that causes cholera. Campylobacters are another example of curved Gram-negative rods. In a Gram film, these often take on the appearance of seagulls as drawn by small children. It must be a coincidence that one species that is a rare human pathogen *Campylobacter lari* is often isolated from seagulls (*larus* is the Latin name for a gull). Related to the campylobacters is *Helicobacter pylori*. This is a spiral organism that is now recognised as the cause stomach ulcers and associated with a number of other medical conditions.

8

Chemotherapy and antibiotic resistance

One of the greatest triumphs of modern medicine has been the introduction of a rational system of antimicrobial chemotherapy to combat infectious diseases. Since time immemorial, folk remedies have exploited moulds or mould extracts to treat infections. In the early days of microbiology, attempts were made to use extracts derived from fungal cultures to prevent surgical wound infection. Joseph Lister used cultures of his own urine to investigate the microbiology of air. He noted that if moulds were present in his cultures, the bacteria that were also there appeared non-motile and degenerate, whereas if bacteria grew without moulds, they were highly motile. Lister concluded that moulds produced a substance or substances that adversely affected the viability of bacteria. He then reasoned that culture filtrates obtained from moulds should prevent infection if used to irrigate surgical wounds. This practice started 60 years before Alexander Fleming described the antibacterial properties of penicillin, produced from a mould that he had originally misidentified.

The problem of producing sufficient antibiotic from mould cultures defeated both Lister and Fleming. Indeed, Fleming was slow to appreciate the clinical applications of his observations. He intended that penicillin should be used as a selective agent in laboratory media rather than to be administered to patients directly. It was not until the early days of the Second World War that an allied Anglo-American effort overcame the problem of large-scale production and penicillin therapy for human infection was properly initiated. By the end of the war penicillin was so plentiful that it was being used to cure cases of gonorrhoea in the allied troops. This was so that they could more quickly be returned to the front line than would otherwise be the case. After the war

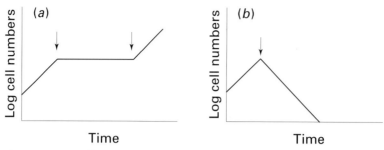

Fig. 8.1. The effect of antimicrobial agents. (a) Addition of a 'static' drug to a broth culture at the time indicated by the first arrow. Cell numbers remain constant until the drug is removed (at the second arrow) and then growth starts once more. (b) Addition of a 'cidal' drug is indicated by the arrow. Cells enter a phase of exponential decline, indicating cell death.

penicillin became generally available, and at one time it was even incorporated into toothpaste in the vain hope it would prevent dental caries. Penicillin was not readily available when the first person was treated: a policeman with overwhelming staphylococcal sepsis. The antibiotic was in such short supply that it had to be re-purified from the patient's urine. Although in this instance the antibiotic caused relief of the clinical condition, once the supply of penicillin had run out, the staphylococcal infection regained its hold and the patient died.

Penicillin represented the first true antibiotic: a substance produced by one microorganism that in very small amounts inhibits or kills other microorganisms. Those agents that kill bacteria are said to be **bactericidal** and those whose effects are reversible upon removal of the drug are **bacteriostatic** (Fig. 8.1). The development of antibiotics was carried out in parallel with the search for chemical antibacterial agents: artificial compounds that inhibit or kill microbes. Paul Ehrlich described such compounds as magic bullets. The most successful of the early antimicrobial compounds, the sulphonamides, are still in use today. Often the term antibiotic is applied very loosely and includes antibacterial agents as well, although this is strictly incorrect.

Bacteria are good targets for the activity of antimicrobial substances. Aspects of their metabolism are significantly different from that of humans. Antibiotics may act upon bacterial reactions that are not found in human cells. This provides the basis for the **selective toxicity** of antibiotics, affecting the bacteria but not the human host. Not all antibiotics are without their side effects. For example, penicillin allergy is very common in humans. The adverse effects of antibiotics are not necessarily associated with their antimicrobial properties. Penicillin allergy results from the presence of the thiazolidine ring of penicillins. It is, however, the β-lactam ring and not the thiazolidine ring that is responsible for the antibiotic activity. Fungi and protozoa have a metabolism that is much closer to that of humans than do bacteria, Moreover, viruses are obligate intracellular parasites that depend almost exclusively upon human metabolism for their replication. As a consequence, antivirus, antifungal and antiprotozoal drugs are more limited in their scope and are generally more toxic to humans than are antibacterial drugs.

8.1 What inhibits bacterial cell wall synthesis?

Peptidoglycan is a macromolecule in which linear polysaccharide chains are extensively cross-linked by short peptide chains. It is a component of bacterial cell walls. It is also an exclusively bacterial polymer and so potentially should provide an excellent target for selective chemotherapy. Unfortunately not all the intermediate steps in peptidoglycan biosynthesis involve processes that are confined to bacteria, and some antimicrobials that inhibit such reactions may be very toxic to humans as well as to bacteria. Peptidoglycan is unique among biological polymers because it contains both L- and D-isomers of its constituent amino acids. Antibiotics may act at several stages during peptidoglycan synthesis. Some antibiotics are valuable chemotherapeutic agents; others are too toxic for human use.

8.1.1 Fosfomycin

Fosfomycin interferes with the condensation reaction between UDP-N-acetylglucosamine and phosphoenol pyruvate in the early stages of peptidoglycan synthesis. There is a rapid selection of resistance to fosfomycin, rendering it unsuitable for most clinical purposes.

8.1.2 Cycloserine

Cycloserine is an analogue of alanine that interferes with two steps in peptidoglycan synthesis. It is a competitive inhibitor of the enzyme that converts L-alanine to D-alanine and it also prevents the action of the enzyme responsible for the formation of D-alanyl D-alanine. The stable ring structure of cycloserine holds the molecule in a sterically favourable position, permitting preferential binding of this compound to both the racemase and the synthetase, rather than their natural substrates. This results in competitive inhibition of these enzymes. Cycloserine is a neurotoxin and is not used clinically except for the treatment of drug-resistant *Mycobacterium tuberculosis,* or in other life-threatening infections where alternative therapies have failed. It is also used in research laboratories to weaken bacterial cell walls prior to lysis.

8.1.3 Bacitracin

Bacitracin is a polypeptide antibiotic that is too toxic for human clinical use. It is, however, widely used in diagnostic laboratories to distinguish bactracin-sensitive *Streptococcus pyogenes* from other β-haemolytic streptococci. Its activity depends upon its ability to bind to the lipid carrier that transports the peptidoglycan monomers across the bacterial membrane. This impedes the dephosphorylation of the carrier, which in turn obstructs regeneration of the carrier and hence prevents re-cycling of the transport mechanism. Bacitracin also interferes with sterol synthesis in mammalian cells by binding to pyrophosphate intermediates, accounting for its human toxicity.

8.1.4 Vancomycin

Vancomycin is a relatively large molecule that acts to prevent the peptidoglycan subunits from being added to the growing cell wall polymer. This is accomplished by vancomycin binding to the D-alanyl D-alanine residue of the lipid-bound precursor. Its primary activity is against Gram-positive bacteria. It is particularly useful in the treatment of serious staphylococcal infections. In these cases it is given either intramuscularly or intravenously since it is not absorbed from the gut. It is also used for the treatment of pseudo-membranous colitis caused by *Clostridium difficile,* when it is administered orally.

8.1.5 Beta-lactams

The β-lactam group of antibiotics includes an enormous diversity of natural and semi-synthetic compounds that inhibit several enzymes associated with the final step of peptidoglycan synthesis. All of this enormous family are derived from a β-lactam structure: a four-membered ring in which the β-lactam bond resembles a peptide bond. The multitude of chemical modifications based on this four-membered ring permits the astonishing array of antibacterial and pharmacological properties within this valuable family of antibiotics. Clinically useful families of β-lactam compounds include the penicillins, cephalosporins, monobactams and carbapenems. Many new variants on the β-lactam theme are currently being explored. Certain β-lactams have limited use directly as therapeutic agents but may be used in combination with other β-lactams to act as β-lactamase inhibitors. Augmentin, for example is a combination of amoxycillin and the β-lactamase inhibitor clavulanic acid. During cross-linking of the peptidoglycan polymer, one D-alanine residue is cleaved from the peptidoglycan precursor and β-lactam drugs prevent this reaction. More recent studies have shown that the activity of this class of drug is more complicated and involves other processes as well as preventing cross-linking of peptidoglycan.

The targets for β-lactam drugs are the **penicillin-binding proteins** (PBPs), so called because they bind radioactive penicillin and can be detected by autoradiography of gels on which bacterial proteins have been separated electrophoretically. The penicillin-binding proteins have transpeptidase or carboxypeptidase activity and they regulate cell size and shape. They are also involved in septum formation and cell division. Bacteria have several individual penicillin-binding proteins, each with a separate function. Conventionally these are numbered according to size, with PBP 1 as the largest protein. The PBP 1 of one bacterium will not necessarily have the same function as the PBP 1 of another.

The β-lactam antibiotics may bind preferentially to different penicillin-binding proteins and at sub-lethal concentrations may cause alterations in cell morphology. For example, mecillinam binds preferentially to *Escherichia coli* PBP 2 and causes spherical cells to form, whereas cephalexin causes *Escherichia coli* to grow as filaments as a result of its preferential binding to PBP 3. This indicates that PBP 2 in *Escherichia coli* is involved in cell elongation whereas its PBP 3 is has a role in the cell division of this bacterium.

The β-lactam antibiotics also stimulate the activity of autolysins. These are enzymes that are responsible for the natural turnover of cell wall polymers to permit growth of the cells. Under normal conditions, these enzymes produce controlled weak points within the peptidoglycan structure to allow for expansion of the cell wall structure. This activity is stimulated by β-lactams, causing a breakdown of peptidoglycan and leading to osmotic fragility of the cell and ultimately to cell lysis.

8.1.6 Isoniazid

Isoniazid inhibits the formation of very-long-chain fatty acids such as those found in the cell walls of mycobacteria. Isoniazid is used in the treatment of tuberculosis and other mycobacterial infections.

8.2 Which antibacterial agents affect bacterial cell membrane function?

Polymyxins and gramicidins act by interfering with the functioning of the bacterial cell membrane by increasing its permeability. Gramicidins are cyclic decapeptides active against Gram-positive bacteria. Polymyxins have a smaller peptide ring attached to a peptide chain ending with a branched fatty acid. They act specifically against Gram-negative bacteria, although chemically modified derivatives do have a broader spectrum of activity. These antibiotics are toxic to humans and are now rarely used in clinical practice. Metronidazole, a very important antianaerobic and antiprotozoal agent probably has a similar mode of action.

8.3 Which antibacterial agents are inhibitors of nucleic acid metabolism?

There are many steps at which nucleic acid metabolism may be interrupted. Antibacterial agents show selective toxicity either because humans lack the metabolic processes that act as targets or because the bacterial targets are much more susceptible to particular chemicals than their eukaryotic counterparts.

8.3.1 Sulphonamides and trimethoprim

Humans are unable to make folic acid, a precursor of purine synthesis. We require an exogenous supply of this metabolite obtained from our diet. Many bacteria are, however, able to generate folic acid from *para*-aminobenzoic acid (PABA) and this pathway provides a target for synthetic antimicrobial agents like the sulphonamides and trimethoprim. Sulphonamides act by inhibition of dihydropteroate synthetase because it acts as a structural analogue of the normal substrate, PABA. Trimethoprim inhibits dihydrofolate reductase, the next step in the folic acid biosynthetic pathway.

Trimethoprim was first introduced to be used in combination with sulphonamides to potentiate their activity. Studies of the combination *in vitro* show that the combination is synergistic. This means that the combined activity of the drugs is more effective than the additive action of the individual components. The synergism observed *in vitro*, however, depends upon maintaining a critical ratio of the two antimicrobials. Because of pharmacological constraints this cannot be achieved in the body, raising doubts about the synergism *in vivo*. Furthermore, using two agents for chemotherapy significantly increases the risk of the patient developing an adverse reaction to the treatment. Such arguments led to the introduction and successful use of trimethoprim as a single agent.

8.3.2 Quinolones

Bacterial DNA exists in a supercoiled form and the enzyme DNA gyrase, a topoisomerase, is responsible for introducing negative supercoils into the structure. Quinolone antibacterial drugs such as nalidixic acid, norfloxacin, ofloxacin and ciprofloxacin act by inhibiting the activity of the bacterial DNA gyrase, preventing the normal functioning of DNA. Humans do possess DNA gyrase but it is structurally distinct from the bacterial enzyme and remains unaffected by the activity of quinolones. These are broad-spectrum agents that rapidly kill bacteria and are well absorbed after oral administration. Overuse of these drugs in certain situations is selecting quinolone-resistant mutants and these may threaten the long-term use of such compounds.

8.4 Which antibacterial agents are inhibitors of RNA metabolism?

Rifampicin inhibits the bacterial DNA-dependent RNA polymerase but has little effect on eukaryotic cells. It is active against the mitochondrial RNA polymerase but its penetration into mitochondria is so poor that it displays very little activity in intact eukaryotic cells. The action of rifampicin prevents production of messenger RNA and this ultimately stops protein synthesis. Clinically, rifampicin is used in treating tuberculosis and for prophylaxis against meningococcal meningitis. In such cases it is offered to close contacts of people with the disease. The synthetic antibacterial nitrofuran compounds also act by preventing messenger RNA production.

8.5   Which antibacterial agents are inhibitors of protein synthesis?

8.5.1   Aminoglycosides

The aminoglycosides are a clinically important group of antibiotics that have a broad-spectrum of activity and are bactericidal in action. The family includes streptomycin, gentamicin, tobramycin, kanamycin, amikacin and netilmicin. The aminocyclitols such as spectinomycin are closely related and have a similar mode of action. Aminoglycosides have a variety of effects within the bacterial cell but principally they inhibit protein synthesis by binding to the 30S ribosomal subunit to prevent the formation of an initiation complex with messenger RNA. They also cause misreading of the messenger RNA message, leading to the production of nonsense peptides. Another important function of the aminoglycosides is that they increase membrane leakage. Antibiotics such as gentamicin and kanamycin exist as mixtures of several closely related structural compounds, those like netilmicin and amikacin have a single molecular structure.

Aminoglycosides are toxic to humans, causing problems with kidney function and damage to the eighth cranial nerve. This leads to hearing loss and balance difficulties. The therapeutic use of the aminoglycosides requires careful monitoring to ensure adequate therapeutic levels are maintained, without the accumulation of the drug to toxic levels.

8.5.2   Tetracyclines

The tetracyclines are a family of antibiotics that have a four-ring structure. They are broad-spectrum agents that inhibit binding of the aminoacyl transfer RNA to the 30S ribosomal subunit in bacteria. The action is bacteriostatic and can be reversed upon removal of the drug. The clinical use of tetracyclines is generally confined to adults. This is because tetracyclines affect bone development and can cause staining of teeth in children.

8.5.3   Chloramphenicol

Chloramphenicol is a broad-spectrum bacteriostatic agent that is toxic to humans. It has been recognised as a cause of aplastic anaemia and so its use is confined to life-threatening infections where no alternative therapy is available. It acts by binding to the 50S ribosomal subunit and blocking the formation of the peptide bond by inhibiting peptidyl transferase activity. It is a potent inhibitor of mitochondrial protein synthesis in eukaryotic cells.

8.5.4 Macrolides

The macrolides are a group of antibiotics that have a large, lactone ring structure. The most widely used macrolide is erythromycin. It is a relatively non-toxic antibiotic, most active against Gram-positive bacteria. Erythromycin is, however, the treatment of choice for Legionnaire's disease caused by the Gram-negative bacillus *Legionella pneumophila* and it is also active against *Haemophilus influenzae*, another Gram-negative bacillus. Erythromycin binds to the 50S ribosomal subunit and inhibits either peptidyl transferase activity or translocation of the growing peptide. Newer macrolides include azithromycin and clarithromycin. These have the same activity as erythromycin but they have better pharmacological properties. The lincosamide antibiotic lincomycin and its semi-synthetic derivative clindamycin have a similar mode of action.

8.5.5 Fusidic acid

Fusidic acid is a steroid antibiotic used to treat Gram-positive infections. It acts by preventing translocation of peptidyl tRNA. Resistant mutants may easily be selected, even during therapy, and therefore fusidic acid is usually administered in combination with another antibiotic. This helps reduce the risk of selecting resistant mutants. To survive, the fusidic acid-resistant mutants must also become resistant to the antibiotic given in combination. If the chance of selecting fusidic acid resistance is 10^{-5} and that of chromosomal mutation to resistance to the second agent is 10^{-8} then the theoretical chance of a double mutant arising from combination therapy is 10^{-13} ($=10^{-5} \times 10^{-8}$). This is so low as to be considered insignificant, but in practice the chance of acquiring resistance to the second agent is increased by the presence of mobile bacterial genes encoding resistance. These can be acquired more easily than chromosomal genes can mutate to confer a resistance phenotype on a bacterium.

8.5.6   Mupirocin

Mupirocin is a molecule that acts as an analogue of isoleucine. It inhibits the isoleucyl-transfer RNA synthetase, thereby preventing the incorporation of isoleucine into growing polypeptide chains. It is not toxic to humans but can only be used topically for skin infections. This is because humans rapidly metabolise the drug to an inactive form. Therefore, in systemic therapy, it is destroyed before it can be effective.

8.6   What drugs act as antifungal agents?

Fungal infections are caused by eukaryotic organisms and for that reason they generally present more difficult therapeutic problems than do bacterial infections. There are relatively few agents that can be used to treat fungal infections. The fungal cell wall may be considered to be a prime target for selectively toxic antifungal agents because of its chitin structure, absent from human cells. No clinically available inhibitor of chitin synthesis analogous to the antibacterial β-lactams exists at present, even though much effort is being directed towards developing such agents. Other targets are currently being exploited.

8.6.1   Polyene antibiotics

Polyene antibiotics bind to sterols within the fungal membrane, disrupting its integrity. This makes the membrane leaky, leading to a loss of small molecules from the fungal cell. Polyene antibiotics include nystatin, used topically for candida infections, and amphotericin B. The latter is administered parenterally and is widely used to treat systemic mycoses. It is most often given intravenously in a bile salt suspension and diluted with 5% dextrose. It penetrates poorly into cerebrospinal fluid and when used to treat meningitis it may be delivered directly into the brain ventricles. Amphotericin B is a very successful and widely used antifungal drug but its use is beset with problems of toxicity. It can cause unpleasant side effects including chills, fever and a lowering of blood pressure. It may also cause kidney damage. The side effects of amphotericin B therapy can mimic the clinical appearance of serious systemic infection, complicating patient management. The severity of side effects may cause interruption of antifungal treatment. Newer lipid-associated forms of amphotericin B are now available. These are considerably less toxic than the older preparations but they are currently very expensive. Amphotericin B remains the drug of choice for life-threatening fungal infections. It may often be administered together with flucytosine since in combination a lower dose may be used, reducing the risk of therapeutic complications.

8.6.2 Azoles

Azoles are an emerging group of antifungal agents that include imidazoles and the more recently developed triazoles. They act to inhibit synthesis of ergosterol, a component of fungal membranes. These drugs, like the polyene antibiotics, may cause leakage of small molecules out of fungal cells. The imidazoles include clotrimazole, miconazole and ketoconazole; the triazoles include fluconazole and itraconazole. They have a broad spectrum of antifungal activity, although there is some variation of activity between the various compounds. Some azoles are also active against Gram-positive bacteria. Azoles are known to exert their action by inhibiting a fungal cytochrome P450 enzyme. As a consequence, some members of the azole group can also affect the human equivalent enzymes and are toxic to humans.

The majority of azoles can only be used topically but some are used to treat systemic infections. Some may be taken orally whereas others must be delivered parenterally. Ketoconazole is administered orally; miconazole is given intravenously. Fluconazole and itraconazole may be delivered either orally or parenterally. These drugs are being evaluated in the treatment of systemic mycoses and have met with variable success. Resistance to fluconazole has emerged during long-term therapy. *Candida glabrata* quickly becomes resistant to this drug. Certain other species, such as *Candida krusei,* are not sensitive to this compound.

8.6.3 Griseofulvin

Griseofulvin is a naturally occurring compound and so is a true antifungal antibiotic. It binds to the proteins involved in microtubule formation and prevents separation of chromosomes at mitosis. Why griseofulvin does not affect human cells is not known. It is used in the treatment of ringworm and other fungal infections of the skin or nails. It has been available since the 1950s but is now becoming superseded by newer drugs.

8.6.4 Flucytosine

Flucytosine, also known as 5-flurocytosine, is a synthetic pyrimidine that interferes with the nucleic acid metabolism in fungi. Because it is an analogue of a naturally occurring nucleotide, it is actively taken up by cells where it is metabolised to 5-fluorouracil. It is a drug used primarily to treat systemic candida infections and it may be given orally or parenterally. It was used as a single agent but is now used exclusively in combination with other antifungal drugs, mainly amphotericin B. Flucytosine achieves good penetration into body fluids, including cerebrospinal fluid. It is excreted through the kidneys and its dosage must be modified in patients with renal problems. It derives its selective toxicity from the inability of human cells to convert it to 5-fluorouracil. Its major drawback is the ease with which resistance develops. Susceptibility testing of fungal isolates is necessary during treatment in order that resistant variants may be detected early and alternative therapy may be initiated.

8.6.4 Allylamines and benzylamines

The allylamines and benzylamines are a new group of synthetic antifungal drugs. Like the azoles they inhibit the synthesis of ergosterol found in the fungal cell membrane. In contrast to the azoles, however, they do not act on a fungal cytochrome P450 enzyme. Consequently, interference with human metabolism is less common and their use is associated with fewer complications. So far the allylamine terbinafine has been the most widely used of this group of drugs. Its principal use has been in the treatment of ringworm infections, particularly those associated with nails. It may be used topically but can also be taken as an oral preparation.

8.7 What drugs can be used to treat virus infections?

Because of the nature of viruses as intracellular parasites, very few clinical useful agents to treat virus infections have been produced. Development of drugs that are active against viruses is one of the most challenging areas in antimicrobial chemotherapy. Drugs that act to treat virus infections generally disrupt the virus lifecycle. It can be difficult to identify targets in the virus lifecycle where disruption of the function will not also damage the infected host. A newly emerging area in the therapy of virus infections is the application of immunomodulating agents. Administration of such drugs enhance the immune response against viruses.

8.7.1 Aciclovir and ganciclovir

Aciclovir or acycloguanosine is a purine. In an infected cell, the thymidine kinase produced by herpesviruses phosphorylates this drug and it is subsequently metabolised into a triphosphate derivative. It is in this form that the drug is active. It enters the nucleoside pool in the cell and competes with guanosine triphosphate. It inhibits the herpesvirus DNA polymerase by incorporation into the replicating virus DNA. Its chemical structure prevents addition of further residues on the virus DNA and, therefore, it is referred to as a chain terminator. It is particularly active against herpes simplex virus types 1 and 2, is less active against varicella zoster virus and has significantly reduced activity against cytomegalovirus. It has prevented significant mortality associated with herpes simplex encephalitis and is of value in treating varicella zoster infections, particularly in immunocompromised patients. It is also used for the treatment of severe cases of genital herpes. Resistance to aciclovir results from mutations in either the thymidine kinase or the DNA polymerase genes. These mutations limit the use of this drug for more trivial conditions. It does not eradicate latent virus but aciclovir does shorten the duration of clinical symptoms. Ganciclovir is an analogue of aciclovir that is active against cytomegalovirus as well as other herpesviruses. It, however, does not prevent chain termination in the growing virus DNA, although it is an inhibitor of the virus DNA polymerase. A number of other nucleoside analogues, each with subtly different properties, are now being marketed.

8.7.2 Amantidine

Amantidine is used to treat shingles, particularly in elderly or debilitated patients. It is also used for the treatment of influenza A virus infections. Its mode of action has not been fully elucidated but it is thought to interfere with the uncoating of the virus. It was thought that this drug worked by raising the pH inside lysosomes. It now appears that it acts by interfering with the penetration of hydrogen ions into the virion. During epidemics of influenza A virus infection, it may be used as a prophylactic agent to prevent disease in the most vulnerable patients. Rimantadine has a similar action.

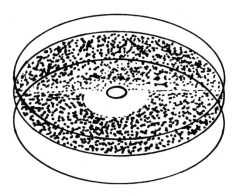

Fig. 8.2. Antimicrobial sensitivity testing. The test plate has been inoculated with two strains and an antibiotic disc has been placed at the interface. The lower strain is sensitive to the antibiotic as shown by the area of inhibition of growth. The upper strain is resistant and can grow up to the disc.

8.7.3   Ribavirin

Ribavirin is a broad-spectrum drug active against both RNA and DNA viruses. In the developed world its principal use is in treating severe respiratory syncytial virus infections in infants. The virus infects the bronchioles, causing bronchiolitis and the drug may be inhaled. Its mode of action is unclear.

8.7.4   Zidovudine

Zidovudine was formerly called azidothymidine and is also known as AZT. It is an antiretrovirus agent that is phosphorylated to form a triphosphate derivative inside infected and uninfected calls. Zidovudine triphosphate is a competitive inhibitor of the retrovirus reverse transcriptase, the enzyme that produces provirus DNA from the virus RNA template. Its principal use is in the management of patients with advanced AIDS. As with aciclovir, there are now a number of reverse transcriptase inhibitors that act like zidovudine.

8.8   What causes antibiotic resistance in bacteria?

Bacteria have evolved numerous strategies for resisting the action of antibiotics and antibacterial agents (Fig. 8.2). This is particularly true of those bacteria that are antibiotic producers. Bacteria that produce antibiotics do so to gain a selective advantage over other competing microbes in their natural environment. If they were sensitive to their own metabolic products, such a selective advantage would be lost. In many hospital units, exploitation of antibiotics is very intensive and this generates an enormous selective pressure for bacteria to acquire the means by which they may become resistant to antibiotics. Under such circumstances it is not unusual to find that bacteria exhibit resistance to more than one group of antibiotics. Resistance to a particular agent may be accomplished by more than one resistance mechanism.

Bacteria may display antibiotic resistance by one or more of the following mechanisms:

- they may lack a target for the antibiotic
- the antibiotic target may be inaccessible
- the antibiotic target may be modified to prevent the action of the drug
- the antibiotic may be chemically modified or destroyed
- bacteria may elaborate alternative pathways, avoiding the drug target.

Not all bacteria have peptidoglycan in their cell wall. Rickettsias and chlamydia, for example, lack peptidoglycan. Such bacteria are intrinsically resistant to the action of cell wall inhibitors such as the penicillins and cephalosporins. Interestingly, although chlamydia do not make peptidoglycan they *do* possess penicillin-binding proteins. This raises the fascinating question: why? What role do these enzymes fulfil in a bacterium that does not have a cell wall?

Having a target that is inaccessible to antibiotics may be achieved in a variety of ways. The outer membrane of Gram-negative bacteria may act as a permeability barrier for antibiotics. Many Gram-negative bacteria are intrinsically resistant to antibiotics like benzylpenicillin because such drugs cannot penetrate the outer membrane and so cannot reach their target. Alterations to the side chain attached to the penicillin nucleus may overcome the problem of membrane penetration and semi-synthetic penicillins such as ampicillin have a broad-spectrum of activity, encompassing both Gram-positive and Gram-negative bacteria. Another way in which bacteria deny access of an antibiotic to its target is actively to pump the drug out of the cell. Gram-negative bacteria may resist the activity of tetracyclines as a result of energy-dependent active efflux of the drug.

Modification of the antibiotic target is often seen in laboratory-generated mutants. For example, resistance to the quinolone antimicrobials results from alteration to the DNA gyrase and aminoglycoside resistance may result from

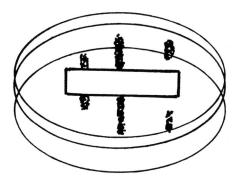

Fig. 8.3. Bacterial interactions with streptomycin. The filter paper laid across these inocula contains streptomycin. The strain on the right is sensitive to the antibiotic and cannot grow near the strip. The strain in the middle is resistant and can grow up to the strip. The strain on the left is dependent upon streptomycin. Its growth is restricted to the region of the plate where the antibiotic has diffused from the strip.

modifications of the ribosome structure. Indeed, ribosomes may be further altered so that they only function in the presence of aminoglycosides. The drug acts to stabilise the functional ribosome in aminoglycoside-dependent bacteria (Fig. 8.3).

Many clinically important bacteria produce enzymes that are capable of chemically modifying or destroying antibiotics. Chloramphenicol may be acetylated by the action of chloramphenicol acetyltransferases. Aminoglycosides may be acetylated by aminoglycoside acetyltransferases, phosphorylated by aminoglycoside phosphotransferases or conjugated with nucleotides. Such modifications render the antibiotic inactive. Antibiotics may also be enzymatically degraded to an inactive form. The β-lactam bond can be hydrolysed by a large family of enzymes known as the β-lactamases. Some β-lactamases have a preferential activity against penicillins and these are referred to as penicillinases. Cephalosporinases are more active against cephalosporins. Recently broad-spectrum β-lactamases have evolved that have activity against both penicillins and cephalosporins. There are families of such enzymes that have arisen as the result of point mutations accumulating in the genes that code for penicillinases. Many of these new enzymes are encoded by self-transmissible plasmids and these new resistance determinants can spread with great ease.

Not all β-lactamase activity is associated with bacterial cells. Human kidney cells produce an analogous enzyme that, although it does not readily attack penicillins and cephalosporins, rapidly destroys carbapenems such as imipenem. Because of this, imipenem is administered together with cilastatin, an inhibitor of the human kidney enzyme. This delays the breakdown of imipenem sufficiently to permit its activity in treating bacterial infection.

Staphylococci have been associated with the production of β-lactamase for many years. Early in the history of the development of semi-synthetic penicillins, compounds were manufactured that were able to resist the activity of staphylococcal penicillinase. These drugs had side chains that prevented the staphylococcal β-lactamase from binding to the antibiotic and hydrolysing it. Methicillin, a staphylococcal β-lactamase-stable penicillin was introduced into clinical practice during the 1960s. Shortly after its introduction, methicillin-resistant strains of *Staphylococcus aureus* (MRSA) were isolated from hospital units where methicillin was in regular use. Methicillin-resistance is greater at 30°C than at 37°C. Resistance results from the temperature-sensitive production of an extra penicillin-binding protein, PBP 2$'$, not susceptible to inhibition by methicillin. Methicillin-resistant *Staphylococcus aureus* also produce a β-lactamase and they are generally resistant to a very wide range of unrelated antimicrobials as a consequence of the expression of an array of resistance genes. Consequently, infections caused by methicillin-resistant *Staphylococcus aureus* can be difficult to treat. In some cases, the only drugs available effectively to treat infections caused by methicillin-resistant *Staphylococcus aureus* are the glycopeptides such as vancomycin. Bacterial resistance to vancomycin was unknown until recently. Vancomycin resistance first appeared in enterococci. These bacteria are resistant to all currently available standard antimicrobial therapies. In an experiment of dubious ethical status, the gene encoding vancomycin resistance was transferred in the laboratory from a vancomycin-resistant enterococcus into a methicillin-resistant *Staphylococcus aureus*. In 1997, the first naturally occurring vancomycin-resistant, methicillin-resistant *Staphylococcus aureus* appeared in Japan. Such bacteria are effectively untreatable at present.

Some antimicrobial resistance genes have only ever been found located on the bacterial chromosome. Others have been found to lie on plasmids. Plasmids encoding antibiotic resistance are often called resistance factors, R-factors or R-plasmids. These plasmids may encode resistance to several unrelated antibiotics. Some R-plasmids are self-transmissible and can move from strain to strain, even between different bacterial genera. Other R-plasmids, although not self-transmissible, may be mobilised by other plasmids that need not necessarily encode antibiotic resistance. Furthermore, antibiotic-resistance genes are frequently located within transposons. These can move more or less at random around the bacterial genome. Subject to certain constraints, genes encoded by transposons may spread very easily because many transposable elements may become associated with transmissible plasmids. In this way, antibiotic-resistance genes may become rapidly disseminated. Resistance genes may even transfer between Gram-positive and Gram-negative bacteria. Tetracycline resistance in the Gram-negative *Neisseria gonorrhoeae* is conferred through expression of a gene that was originally described as part of a transposon isolated in the Gram-positive streptococci. During the 1980s, this gene became very widely disseminated throughout many bacterial species. Control of antimicrobial resistance in pathogenic microbes is one of the greatest challenges currently facing medical microbiology. If we are unsuccessful, we will surely enter the post-antibiotic era.

Further reading

The interested reader may wish to consult the following texts in their future studies.

Bergey's Manual of Systematic Bacteriology: Vol. I Kreig, N.R., ed. (1984); Vol. II, Sneath, P.H.A., ed. (1986); Vol. III Stanley, J.T., ed. (1989); Vol. IV, Williams, S.T., ed. (1989). Baltimore, MD: Williams & Wilkins.

Dale, J.W. (1994). *Molecular Genetics of Bacteria*, 2nd edn. Chichester, UK: Wiley.

Deacon, J.W. (1997). *Introduction to Modern Mycology*, 3rd edn. Oxford: Blackwell Scientific.

Department of Health (1978). *Code of Practice for the Prevention of Infection in Clinical Laboratories and Post-mortem Rooms*. London: Her Majesty's Stationary Office.

Dimmock N.J. & Primrose, S.B. (1994). *Introduction to Modern Virology*, 4th edn. Oxford: Blackwell Scientific.

Dixon, B. (1994). *Power Unseen: How Microbes Rule the World*. London: Freeman.

Ellner, P.D. & Neu H.C. (1992). *Understanding Infectious Disease*. London: Mosby.

Emmerson, A.M., Hawkey, P.M. & Gillespie S.H. (1997). *Principles and Practice of Clinical Bacteriology*. Chichester, UK: Wiley.

Gerhardt, P., Murray, R.G.E., Costilow, R.N., Nester, E.W., Wood, W.A., Kreig, N.R. & Phillips, G.B. (eds.) (1994). *Manual Methods for General and Molecular Bacteriology*. Washington DC: American Society for General Microbiology.

Heritage, J., Evans E.G.V. & Killington, R.A. (1996). *Introductory Microbiology*. Cambridge: Cambridge University Press.

Humphries, H. & Irving, W.L. (1996). *Problem-orientated Clinical Microbiology and Infection*. Edinburgh: Churchill Livingstone

Lim, D.V. (1989). *Microbiology*. St Paul, MN: West Publishing.

Madigan, M.T., Martinko, J.M. & Parker, J. (1997) *Brock Biology of Microorganisms*, 8th edn. New York: Prentice Hall.

Mims, C.A. (1987). *The Pathogenesis of Infectious Diseases*, 3rd edn. New York: Academic Press.

Mims, C.A., Playfair, J.H.L., Roitt, I.M. Wakelin, D. & Williams, R. (1998). *Medical Microbiology*, 2nd edn. London: Mosby.

Mims, C.A., Urwin, G. Zuckerman, M. (1994). *Case Studies in Medical Microbiology*. London: Mosby.

Nester, E.W., Roberts, C.E. & Nester, M.T. (1995). *Microbiology, a Human Perspective*. Oxford, UK: William C. Brown.

Postgate J. (1992). *Microbes and Man*, 3rd edn. Cambridge: Cambridge University Press.

Postgate J. (1994). *The Outer Reaches of Life*. Cambridge: Cambridge University Press.

Prescott, L.M., Harley, J.P. & Klein, D.A. (1996). *Microbiology*, 3rd edn. Oxford, UK: William C. Brown.

Russell, A.D. & Chopra, I. (1996). *Understanding Antibacterial Action and Resistance*, 2nd edn. London: Ellis Horwood.

Schaechter, M., Medoff, G. & Schlessinger, D. (1989). *Mechanisms of Microbial Disease*. Baltimore, MD: Williams & Wilkins.

Schlegel, H.G. (1993). *General Microbiology*, 7th edn. Cambridge: Cambridge University Press.

Schulman S.T., Phair, J.P. & Sommers, H.M. (1992). *The Biologic and Clinical Basis of Infectious Diseases*, 4th edn. Philadelphia, PA: W.B. Saunders.

For those with access to the World Wide Web we maintain a site that links to some important microbiology resources. The URL is:

http://www.leeds.ac.uk/mbiology/fullproj/proj1.htm

Glossary

Abortive poliomyelitis   Term sometimes used to describe infections caused by the poliovirus in which the patient does not develop the paralysis or muscle-wasting characteristic of the disease.

Abscess   Defined localised lesions within a tissue. Often abscesses are polymicrobial: that is they are caused by more than one microbe.

Actinorrhiza   Mutual association of actinomycete bacteria and plant roots. Actinorrhizas may play a role in nitrogen fixation.

Activated sludge tanks   Sewage treatment tanks in which material is vigorously aerated to permit the digestion of waste matter.

Aerobic bacteria   Those bacteria that require oxygen for growth.

Algal bloom   Overgrowth of cyanobacteria and algae in water that becomes nutrient rich. In excessive quantities, these can be toxic. Cattle and dogs have died after drinking from such overgrown waters.

Alkaloid   A nitrogenous organic chemical derived from plants or fungi that may have medicinal properties when taken in small quantities. In large doses alkaloids are usually poisonous. Examples include atropine, muscarine, quinine and strychnine.

Anaerobic bacteria   Those bacteria that cannot grow or are killed by the presence of oxygen. Facultative anaerobes can grow in the presence or absence of oxygen.

Anaphylaxis   Type I hypersensitivity. This is associated with conditions such as hay fever and asthma.

Anorexia   Lack of appetite.

Antibody   A type of protein produced by vertebrate animals in response to the presence of a foreign antigen and that binds in a highly specific manner to that antigen.

Antigen   A substance that elicits the production of antibodies. Antigens are usually carbohydrates or proteinaceous in nature.

Aplastic anaemia A form of anaemia in which the cells of the bone marrow responsible for blood formation become suppressed and blood cells can no longer be made. Although this can have several causes, it is one of the serious and life-threatening side effects of the use of chloramphenicol.

Aquifer Rock or soil that holds a considerable quantity of water. Aquifers may be tapped for drinking supplies.

Arbuscule Feathery structure formed by a fungus inside the plant root cells in an endomycorrhiza.

Archaebacteria A large group of bacteria, some of which can live in extremely hostile environments. They were amongst the first living organisms.

Arthritis Inflammation or infection of a bone joint.

Aseptic meningitis A term used to describe virus meningitis. Literally, a meningitis in which the cerebrospinal fluid does not yield a positive microbial culture.

Aseptic technique Manipulative techniques designed to reduce to a minimum the risks of contaminating either cultures or the environment.

Assimilatory nitrate reduction The incorporation of ammonia into organic polymers.

Asymptomatic bacteriuria The presence of bacteria in urine without clinical signs of infection.

Attenuation The process whereby microbes are treated so that they can no longer cause overt disease, yet they remain able to cause infection.

Autolytic enzymes Those enzymes that cause destruction of the tissue or organism that produced them.

Auxotrophic mutant A mutant that, in order to grow, requires nutritional substances not required by wild-type strains.

Bacteraemia The presence of bacteria in the bloodstream.

Bactericidal agent An agent that kills bacteria.

Bacteriostatic agent An agent that prevents bacterial growth.

Bacterium (plural: bacteria) A prokaryotic microorganism.

Bacteroids Degenerate and irregularly shaped cells of *Rhizobium* species found within root nodules.

Balanitis Inflammation or infection of the penis.

Biogenic Produced by living organisms.

Biological filters Large tanks used in the treatment of sewage and relying upon the activity of microbes in the complex ecosystems that develop in filter beds.

Biological oxygen demand (BOD) The quantity of dissolved oxygen consumed by a defined volume of waste material held in the dark at a specified temperature and for a specified time. Typically, one litre of waste is held in the dark at 20°C for five days in the measurement of the BOD.

Biopolymers Large molecules made by living organisms and made up from repeating monomeric subunits. Proteins and nucleic acids are good examples.

Bioremediation The use of microbes to reclaim natural ecosystems following oil or chemical spills.

Blanching Culinary term for the brief immersion of food in boiling water, followed by rapid cooling. The process does not cook the food thoroughly but will cause disruption of biopolymers thereby preventing the activity of autolytic enzymes.

Buboe Enlarged, infected lymph node, typically seen in bubonic plague but also a clinical feature of the sexually transmissible disease lymphogranuloma venereum.

Calculus (plural: calculi) Stones formed in tissues.

Candidosis Infection caused by *Candida* species. Oral and vaginal candidosis is commonly called 'thrush'.

Carbuncle A collection of hair follicles infected with *Staphylococcus aureus* causing a crop of boils.

Caries Decay of hard tissues such as teeth or bone.

Carriers Individuals who harbour pathogens without showing any evidence of disease.

Cell-mediated immunity The immune response involving the activity of non-antibody producing cells such as T cells.

Cellulitis Rapidly spreading skin infection caused by *Streptococcus pyogenes* and, occasionally, by other bacteria.

Cerebrospinal fluid Clear, colourless fluid found in the ventricles of the brain and in the spinal cord. This fluid undergoes characteristic pathological changes in patients suffering meningitis.

Chancre Ulcerated lesion associated with sexually transmissible diseases. Hard chancres are a feature of primary syphilis, caused by *Treponema pallidum*, and soft chancres appear in chancroid, a disease caused by *Haemophilus ducreyi*.

Chemolithotroph An organism that oxidises inorganic compounds to gain energy and uses carbon dioxide as its sole carbon source. Chemolithotrophs do not require light.

Chemoorganotroph An organism that uses organic compounds as a source of energy and that does not require light to grow.

Choleragen Cholera toxin, responsible for causing a profuse, watery diarrhoea.

Commensal flora The normal microbial flora found in association with higher organisms.

Conjunctivitis Inflammation or infection of the eyes.

Controlled atmospheric packaging (CAP) Monitoring and manipulation of the atmosphere over foods to prevent the growth of certain spoilage organisms. CAP involves periodic manipulation of the atmosphere. *See also* modified atmospheric packaging.

CSU Catheter specimen of urine. These are collected from the bag attached to the end of a catheter: a tube passed up the urethra and into the bladder.

Cyanosis A bluish appearance of the skin and mucous membranes caused by poor oxygenation of the underlying tissues.

Cystitis Inflammation or infection of the bladder.

Demyelination Removal of the myelin sheath that surrounds nerves. This leads to impaired nerve conduction and neurological disease.

Diarrhoea   Excessive production of fluid rather than solid faeces.

Dichotomous   Divided into two.

Differential coliform count   Measure of the number of coliform bacteria originating from faecal material.

Dimorphic fungus   A fungus that lives part of the time as a mould and at other times exists in a yeast form.

Disseminated intravascular coagulation   Clotting of blood throughout the circulatory system. This often results from endotoxic shock and causes multiple organ failure as tissues are deprived of oxygen. It is often a fatal condition and is sometimes referred to as 'DIC' or 'Death is Coming'.

Dissimilatory nitrate reduction   The conversion of organic nitrogen compounds to an inorganic form.

Dissimilatory sulphur metabolism   Conversion of organic sulphur compounds into inorganic compounds such as hydrogen sulphide: rotten egg gas.

Dysuria   Pain when passing urine.

Ectomycorrhiza   Mycorrhiza in which the fungus merely surrounds the root, with no penetration of plant tissue.

ELISA   Enzyme-linked immunosorbent assay. This technique involves attaching an enzyme to a desired antibody. The antibody then reacts with a fixed antigen and a colourless substrate is added to the reaction. If the enzyme-linked antibody can react with the fixed antigen, then the enzyme will produce a coloured product. In a variation of the ELISA test, the enzyme may be attached to a soluble antigen and this complex reacted with an immobilised antibody.

Embolus (plural: emboli)   Matter carried around the bloodstream that causes blockage of a blood vessel, derived from the Latin word *embolismus* meaning a peg or stopper.

Encephalitis   Inflammation or infection of the substance of the brain.

Endemic   An infection constantly present in a community.

Endocarditis   Infection of the lining of the heart, most frequently affecting previously damaged heart valves.

Endogenous infection   Infection acquired from the patient's commensal flora.

Endometriosis   Inflammation or infection of the uterus.

Endomycorrhiza   Mycorrhiza in which the fungus partner penetrates the root tissue. This may be superficial penetration at one extreme. At the other, the fungus penetrates plant cells.

Endosymbiont   An organism that lives entirely within another organism and where both partners derive benefit from the arrangement.

Endotoxic shock   Fever, increased rate of breathing and heart beat and lowered blood pressure resulting from exposure to endotoxin associated with the outer membrane of Gram-negative bacteria.

Endotoxin   Lipopolysaccharide. This is associated with the outer layer of the outer membrane of Gram-negative bacteria.

Enteritis   Inflammation or infection of the gut.

Epidemic   The occurrence of an infection in numbers significantly above the endemic level.

Epitope   The region of an antigen that stimulates the formation of antibody and to which the antibody binds.

Erysipelas   Cellulitis spreading over the face.

Eukaryote   An organism in which the cell or cells have a nucleus separated from the cytoplasm by a nuclear membrane and a cytoplasm that contains numerous organelles together with an elaborate arrangement of intracellular membranes.

Exogenous infection   Infection acquired from a source other than the patient's commensal flora.

Facultative anaerobes   Organisms that can grow in the presence or absence of oxygen.

Faecal indicator organisms   Organisms that can typically be found in the guts of animals and that are used as indicators of faecal contamination of water.

Fermentation   Anaerobic breakdown of organic compounds to produce alternative organic compounds. Perhaps the most familiar is the fermentation of glucose by yeasts to produce ethanol.

Flatulence   The production of gas, or flatus, within the alimentary canal. These gasses may be expelled from either end of the gut.

Fomites   Inanimate objects that act as vectors, spreading infection.

Food poisoning   A special case of food spoilage in which apparently wholesome food will cause illness when eaten.

Food spoilage   The processes that render a food inedible. This may be for aesthetic reasons as well as for reasons of safety.

Fungaemia   The presence of fungi in the bloodstream.

Fungus (plural: fungi)   A eukaryotic organism that possesses a cell wall and that requires a supply of organic matter from which it derives its energy.

Furuncle   A boil, resulting from the infection of a single hair follicle with *Staphylococcus aureus*.

Gangrene   Anaerobic, highly destructive infection of soft tissues most often caused by *Clostridium perfringens*.

Genetic drift (of influenza virus)   The accumulation of point mutations, which alter subtly the antigenic structure of the virus.

Genetic shift (of influenza virus)   The re-assortment of virus RNA molecules within a virus particle, resulting from double infection of a single cell. Genetic shift is a rare event and results in major antigenic changes in the virus.

Geosmin   Secondary metabolic product of streptomycete bacteria that gives soil its characteristic smell, especially apparent immediately after digging.

Germ tube   Pseudomycelium produced by cells of *Candida albicans* when incubated in serum at 37° C for two hours.

Gingivitis   Infection or inflammation of the gums.

Glomerulonephritis Inflammation of the kidney particularly affecting the glomeruli. This damage is caused by circulating immune complexes and is a rare complication of streptococcal infection.

Gumma Ulcerated lesions associated with tertiary syphilis.

Gummosis The extrusion of gums from infected plant tissues.

Haematogenous spread Spread (of microbes) through the bloodstream.

Haematuria The presence of blood in urine.

Haemorrhage Escape of blood from blood vessels: profuse bleeding.

Halophilic Literally, salt-loving. Halophiles can tolerate high salt concentrations. Extreme halophiles can live in very highly salty environments such as the Great Salt Lake in Utah or the Dead Sea in the Middle East.

Haustorium (plural: haustoria) Specialised branching structure produced by pathogenic fungi to penetrate plant cells and through which they obtain food.

Hermetic Air tight

Heterocysts Specialised non-photosynthetic cyanobacterial cells responsible for nitrogen fixation.

Heterofermentative bacteria Those bacteria that produce a variety of organic compounds through fermentation reactions.

Heterophile antibodies Those antibodies that react differently to different antigens. The presence of heterophile antibodies is a traditional marker of Epstein–Barr virus infection causing glandular fever.

Heterotroph An organism that requires one or more organic compound to grow.

Homofermentative bacteria Those bacteria that produce a single organic compound through fermentation reactions.

Hybridoma Cell formed from the fusion of a myeloma cell and an antibody-producing spleen cell and used as the source of monoclonal antibody.

Hydrocephalus Excessive accumulation of cerebrospinal fluid in the ventricles of the brain.

Hypersensitivity Overactive immune response seen on second or subsequent exposure to an antigen. There are four major types of hypersensitivity reaction.

Hypothermia Lowered body temperature.

Immunoglobulin A glycoprotein that functions as an antibody. There are several classes of immunoglobulin, each with a separate function.

Impetigo A superficial bacterial skin infection associated with pus production and leading to encrusted lesions.

IMViC tests Group of tests including indole production, the Methyl Red and Voges–Proskauer tests and the utilisation of citrate as a sole carbon source. Together these tests can be used in the presumptive identification of coliform bacteria.

Infection thread Structure formed during the development of root nodules in legumes.

Isomer One of two or more compounds for which the molecular formula is the same but the arrangement of atoms is different. This difference alters the reactivity of the various isomers, particularly with respect to enzyme reactions. Isomers often have mirror-image structures.

Kaposi's sarcoma   Malignant tumour associated with AIDS patients.

Koch's Postulates   The criteria used to determine whether a microbe is the cause of a particular infection.

Lactose   Sugar found in milk. Lactose fermentation is an important characteristic used in the identification of faecal pathogens.

Leavened bread   The bread made with yeast to make the dough rise; *Levare* is the Latin verb to lift or raise.

Leghaemoglobin   Form of haemoglobin found in the root nodules of nitrogen-fixing plants and responsible for their pink colour.

Lepromatous leprosy   Chronic skin infection caused by *Mycobacterium leprae* in which the patient exhibits little cellular immune response.

Limiting nutrient   A nutrient the concentration of which regulates the growth of microbes in a given habitat.

Lumbar puncture   Procedure in which a needle is introduced into the spinal cord between the vertebrae of the lumbar spine. This enables withdrawal of a sample of cerebrospinal fluid.

Lyophilisation   Freeze-drying.

Lymphogranuloma venereum   Rare and serious venereal infection caused by *Chlamydia trachomatis*.

Malaise   Generalised, unfocused bodily discomfort: feeling awful.

Mashing   The process whereby malt is extracted from grain at the beginning of the brewing process.

Meningitis   Inflammation or infection of the meninges: membranes surrounding the brain.

Microaerophile   An organism that is damaged by atmospheric concentrations of oxygen but that requires some oxygen, typically 2–5%, to be present to grow.

Modified atmospheric packaging (MAP)   Alteration of the gasses over packed food. The technique was developed to suppress the growth of spoilage organisms. MAP involves a single manipulation step in contrast to controlled atmospheric packaging (CAP) in which the atmosphere is monitored and manipulated as necessary.

Monoclonal antibody   An antibody of a single type that is produced by fusing an antibody-forming cell with a type of cancer cell and maintaining the resultant hybrid in artificial culture.

Mould   A filamentous fungus.

MSU   Mid-stream specimen of urine.

Must   Fruit pulp used as the starting material in the production of wine.

Myalgia   Muscle pain.

Mycelium (plural: mycelia)   The network of filaments that constitutes the vegetative structure of moulds and streptomycete bacteria.

Mycorrhiza   Mutualistic relationship between fungi and the roots of a plant.

Mycosis (plural: mycoses)   General term for a fungal infection.

Myeloma cell   A cell derived from a bone marrow tumour. When fused with antibody-producing cells the resulting hybrid cell produces large quantities of monoclonal antibody.

Myocarditis   Infection of the heart muscle.

Necrosis   Death of a defined area of tissue.

Necrotising fasciitis   Aggressive bacterial infection of the skin and underlying tissues carrying a high risk of mortality.

Negri bodies   Characteristic cytoplasmic inclusions within the brain seen in cells of the hippocampus of an individual who has died of rabies.

Nitrifying bacteria   Those bacteria that are responsible for the oxidation of ammonia.

Nitrogenase   The enzyme responsible for nitrogen fixation.

Nitrogen assimilation   The uptake of inorganic nitrogen compounds into organic compounds.

Nitrogen fixation   The conversion of nitrogen gas into ammonia, making nitrogen available for uptake into organic compounds.

Nosocomial infection   An infection acquired in hospital.

Obligate aerobes   Organisms that cannot grow without oxygen.

Obligate anaerobes   Organisms that cannot grow or that are killed in the presence of oxygen.

Obligate intracellular parasite   An organism that can only replicate inside the cell of a host organism.

Obligate parasite   An organism that can only replicate within a host organism.

Oedema   Collection of fluid in a tissue causing it to swell. The swelling of oedema is soft and, if depressed, the tissue takes some time to resume its natural shape.

Ophthalmia neonatorum   Eye infection in babies born to mothers infected with *Neisseria gonorrhoeae*. This may lead to blindness if the condition is left untreated.

Opportunistic infections   The infections in compromised individuals caused by pathogens that do not damage healthy individuals.

Opportunistic pathogen   A pathogen that causes disease in compromised hosts but not in healthy individuals.

Opsonin   A substance that makes material more amenable to phagocytosis. Complement and various antibodies act as good opsonins.

Orchitis   Inflammation of the testes. *Orchio* is Greek for testis and it is also from this root that the word 'orchid' is derived.

Osteomyelitis   Infection of the bone.

Oxidation ponds   Small-scale sewage treatment plants relying upon the oxidation of organic material.

Pandemic disease   A disease that is active around the world.

Parasite   An organism that derives its nutrients from a living plant or animal, often but not always to the detriment of its host.

Parasitaemia   The presence of parasites in the bloodstream.

Parenteral (of a drug) Administered by a route other than orally. This may be intramuscular, into a muscle; intravenous, through a vein; or intrathecally, into the ventricles of the brain.

Pasteurisation Heating of a substance to kill vegetative microbes. This is often used to kill spoilage organisms and pathogens in milk to render the product safe to drink and to extend its shelf life.

Virulence factor A structure or attribute of a microbe that enhances its capacity to produce symptoms of disease in an infected host.

Pathogens Microbes that cause infectious disease.

Penicillin-binding proteins Enzymes responsible for the formation and maintenance of the peptidoglycan cell wall of bacteria. These enzymes are the target for β-lactam antibiotics including the penicillins. These drugs bind tightly to penicillin-binding proteins.

Periodontal disease Inflammation or infection of the tissues surrounding a tooth.

Photolithotrophs Organisms that uses light energy and inorganic compounds for growth and that use carbon dioxide as their sole carbon source.

Photoorganotroph An organism that uses light energy and requires a supply of organic compounds for growth.

Photophobia Inability to tolerate bright light.

Plasmolysis Shrinkage of the cell protoplast away from the cell wall because of osmotic removal of water from the cell.

Poliomyelitis Infection caused by the poliovirus that causes paralysis and muscle wasting.

Polyclonal antibody An antibody produced by the normal immunisation procedure. Serum containing polyclonal antibodies will have one principal antibody but this will be mixed with other minor antibodies each capable of reacting with different epitopes.

Polymerase chain reaction (PCR) The reaction used to amplify a sequence of DNA lying between two target sequences. Oligonucleotide primer pairs are designed to flank the desired sequence, one for each DNA strand. There then follows repeated cycles of strand separation, primer annealing and DNA replication to yield the DNA product known as an amplimer.

Polymorphonuclear leukocytes Short-lived white blood cells that provide a first line of defence against bacterial infection. These cells are the most common of the circulating leukocytes and, because of their staining properties, they are also known as *neutrophils*.

Potable (of water) Safe to drink.

Pre-icteric Phase of hepatitis before the patient develops jaundice.

Presumptive coliform count Measure of the total number of coliform bacteria in a sample. These may or may not originate from faecal contamination. The differential coliform count is used to enumerate coliforms that do have a faecal origin.

Prion An infectious particle thought to cause diseases such as scrapie in sheep, bovine spongiform encephalopathy in cattle and Creutzfeldt–Jakob disease in humans. They have a protein structure but have not been found to contain nucleic acid.

Proctitis   Inflammation or infection of the rectum.

Prognosis   Predicted outcome of a disease process.

Progressive multifocal leukoencephalopathy   Neurological disease associated with the papovavirus JC in which numerous white plaques of demyelination occur throughout the brain. This leads to increasing neurological damage and ultimately death.

Prokaryote   An organism that lacks a membrane-bound nucleus and that has a relatively simple cell architecture.

Prophylactic therapy   Preventative treatment.

Proteolysis   The metabolic breakdown of proteins.

Pseudosepticaemia   A condition in which *after collection* a blood culture becomes contaminated with a bacterium that may cause septicaemia while the patient remains uninfected.

Pus   Thick, yellowish fluid produced in response to certain infections, containing serum, leukocytes and dead cells.

Putrefaction   The process of decay that leads to a loss of structure. Rotting.

Pyelonephritis   Infection of the kidney.

Pyogenic cocci   A name used to refer to *Staphylococcus aureus* and *Streptococcus pyogenes* because of their ability to induce pus formation. *Puon* is the Greek word for pus.

Pyrexia   Raised body temperature. *Pyros* is the Greek word for fire.

Pyuria   The presence of pus in urine.

Quiescent   Dormant: inactive.

Racking   Removing the clear fermented liquid from the sediment produced during winemaking.

Reagin   Complement-fixing substance in the blood of patients infected with syphilis.

Reiter's syndrome   Rare complication of chlamydial and other infections, characterised by arthritis and conjunctivitis.

Reservoir of infection   The principal habitat from which an infectious agent may spread.

Resident flora   Alternative term for the commensal flora.

Respiration   An energy-yielding process involving the oxidation of a substrate such as glucose.

Reverse transcriptase   The enzyme used to make a DNA copy from an RNA template.

Rheumatic fever   An autoimmune disease and complication of streptococcal infection associated with acute swelling and pain affecting one or more joints causing stiffness. Rheumatic fever may cause damage to heart tissues, pre-disposing the patient to endocarditis.

Rhizosphere effect   The accumulation of microbes around the roots of plants.

Ringworm   Infection of the skin that, despite its name, is caused by a fungus.

Salpingitis   Inflammation or infection of the fallopian tubes.

Saprophyte   An organism that lives on dead or decaying organic matter.

Schmutzdeke   Slime layer containing a complex community of microbes that develops on the surface of filter beds. An active schmutzdeke is important in the purification of drinking water.

Selective toxicity   The idea that a compound will have a much greater effect on one organism than another. Selective toxicity can be used to explain why antibiotics can kill microbes and yet not affect humans.

Septicaemia   The presence of microbes in the bloodstream causing symptoms of disease. This is a life-threatening condition.

Septic tank   Large tank designed to contain sewage in which solid material sinks to the bottom where anaerobic digestion reduces the organic content of the effluent.

Serovar   Strain differentiated from others by its antigenic structure and so-called because it is identified in reactions using specific antiserum.

Shiga toxin   The toxin responsible for the mucosal damage and diarrhoea of bacilliary dysentery.

Silage   Cattle feed made by the anaerobic breakdown of vegetable matter.

Source of infection   The location from which an infection is acquired.

Sporadic (of infection)   The occurrence of isolated and apparently unrelated cases.

Sporophores   Spore-bearing structures produced by fungi. In the case of basidiomycete fungi, the sporophores are more commonly referred to as mushrooms or toadstools.

Stormy clot   The reaction of *Clostridium perfringens* in milk. The milk becomes clotted and so much gas is produced that large rips appear in the clotted milk, as the bubbles rise rapidly through the milk.

Stromatolites   Mats of microorganisms often now made from filamentous cyanobacteria. These can become fossilised.

Subunit vaccine   A nucleic acid-free vaccine that is made from those antigenic components of a pathogen that can elicit a protective immune response.

Synergy   When the combined effect of drugs is greater than the sum of their individual activities.

Systemic lupus erythematosus (SLE)   Autoimmune disease that predominantly affects young women. Patients may present with a variety of problems including fever, arthritis, renal and neurological problems. Patients with SLE may yield false-positive results in syphilis serology.

Tabes dorsalis   Neurological features of tertiary syphilis.

Thermoacidophile   An organism that lives at high temperatures and under very acid conditions.

Thermophile   An organism that lives at high temperatures.

Thrombosis   A clot of blood within a blood vessel. This may cause blockage of the vessel, leading to damage in the tissue supplied by the affected vessel.

Toxoid   A treated toxin that retains its immunological structure but which no longer causes damage to tissues.

Transient flora   Microbes that colonise a site for a short time but are easily removed.

Tuberculous leprosy   Chronic skin infection caused by *Mycobacterium leprae* in which the patient exhibits a strong cellular immune response.

Urease   Enzyme responsible for the liberation of ammonia, carbon dioxide and water from urea. This causes a local rise in pH.

Uropathogenic   Capable of causing infection in the urinary tract.

Vacuum-packaging   The packaging of food to remove air. This will remove oxygen and prevent growth of obligate aerobic spoilage organisms such as the pseudomonads. Vacuum-packed food may still support the growth of anaerobic bacteria, including *Clostridium perfringens* and *Clostridium botulinum*.

Vectors of infection   Objects used to spread infection. These may be animate (rat fleas that spread plague) or inanimate. Inanimate vectors of infection are referred to as fomites.

Vegetation   Growth of bacteria and fibrin deposited on an infected heart valve in endocarditis.

Viraemia   The presence of viruses in the bloodstream.

Viroid   Naked RNA molecule that can infect plants without the assistance of any protein.

Virulence   The measure of pathogenicity or the capacity of an organism to cause disease. A virulence factor is an attribute that enables a pathogen to cause disease.

Virus   An infectious particle comprising a nucleic acid, either RNA or DNA, and a protein coat. Some viruses also have a lipid-containing envelope. All viruses are obligate intracellular parasites.

Vitamin   Organic compound essential in trace amounts for the health of an organism but which must be obtained from its environment. Certain vitamins required by animals may be supplied by bacteria of their commensal flora.

Western blotting   The technique of antigen detection in which proteins are separated electrophoretically, transferred to a membrane support and are then exposed to labelled antibody. The desired antigens may then be detected after the membrane has been thoroughly washed to remove unbound antibody.

Whitlow   An infected fingernail.

Wort   The malt liquid product of the mashing process used to brew ales and lagers.

Xenobiotic compound   A compound that is entirely artificial and that does not naturally occur on Earth.

Yeast   A predominantly unicellular fungus.

Index